Other books by Gary Griggs

The Earth and Land Use Planning
with John Gilchrist

Geologic Hazards, Resources and Environmental Planning
with John Gilchrist

Living with the California Coast, author and editor
with Lauret Savoy

Living with the Changing California Coast, author and editor
with Lauret Savoy and Kiki Patsch

Santa Cruz Coast: Then and Now
with Deepika Shrestha Ross

Introduction to California's Beaches and Coast

California Coast from the Air: Images of a Changing Landscape
with Deepika Shrestha Ross

Our Ocean Backyard

Collected Essays

Gary Griggs

Copyright © 2014 Gary Griggs

All rights reserved. No part of this book may be reproduced in any form or by any means without permission in writing from the author.

Colophon
Cover and Book Design by Deepika Shrestha Ross | DSR Studio, using Adobe® InDesign® and typefaces Optima, Franklin Gothic Book and Franklin Gothic Medium.

Cover photographs:
Remains of Pier at Moss Landing, copyright © 2011 D Shrestha Ross
Gary Griggs at Point Lobos, copyright © 2008 D Shrestha Ross

DEDICATION

This book of stories is dedicated to my three grandchildren - Parker in Santa Cruz, and Dylan and Sofia in Barcelona - with the promise that I will do all that I can to ensure that they have a healthy ocean to enjoy, and a sustainable planet to live upon.

Contents

Dedication iii
Acknowledgements xiii
Foreward by Sam Farr xv
Foreward by John Laird xvii
Preface OUR OCEAN BACKYARD - The First Six Years xix

PART I: History of the Monterey Bay Region
Exploring the North Coast 2
Forty Miles of Bad Road 5
The Old Coast Road 8
Hydrocarbons in the Hills 11
Ships of Cement 14
Shipwrecks in the Sanctuary 18
California Whaling 21
Whaling in the Bay 25
West Cliff - A Local Treasure 29
View of the Santa Cruz Waterfront - 1906 32
The Wharves of Santa Cruz 35
Finding Monterey Bay 39
Coastal Place Names 44
Exploring the North Coast - Names & Places 46
Captain Davenport's Landing 49
Yellow Bank Beach 52
Sand Hill Bluff 55
Early North Coast Ranchers 58
Wilder Ranch Still Wild 61
Beach Trash or Beach Glass? 64

PART II: Geology & Field Trips

Peeling Back the Layers ... 68
   Living on an Ancient Seafloor ... 71
   Our Flooded Edge ... 73
      Sand Dunes - Blowing in the Wind ... 75
      Natural Curves Along the Shoreline ... 78
      Submarine Canyons - Sediment Highways
         to the Deep Sea ... 81
      Grand Canyons on the Seafloor ... 84
      Submarine Canyons - Going Deeper ... 86
      Why Submarine Canyons? ... 89
      Why Monterey Submarine Canyon? ... 91
      Ancient Mud, Diatoms, and Whales ... 94
      A Walk to Remember ... 97
      Walking Around the Bay ... 100
      Walking Around the Bay - Again ... 103
      On the Beach - Moss Landing to Monterey ... 107
      Walking on Rocks ... 111
      Wilder's Fallen Arch ... 114
      West Cliff - Stepping Back in Time ... 117
      The Purisima - From Clams to Cetaceans ... 120
      The California Coastal Records Project ... 123

Part III: Coastal Erosion, Protection & Shoreline Change

   California Coast - Worn at the Edges ... 126
   Broken Bridges and Fallen Arches ... 129
   Migrating Shorelines ... 132
   Hazards of Living on the Edge ... 135
   Dealing with a Retreating Coastline ... 137
   Perils of Paradise - Living at the Edge ... 140
   Seacliffs and Seawalls ... 143
   Armoring the Coast ... 146
   Return to Pleasure Point ... 149
   Protecting Pleasure Point ... 152

Coastal cliffs and Rolling Rocks..................... 155
The Ever Changing Coast............................ 158
Coastal Promontories................................ 162
Untrained River Mouths............................. 166
Meandering River Mouths........................... 170
A Castle on the Beach............................... 173
Walls Around our Coastal Cities?.................... 176

Part IV: Beaches
Why Are Our Beaches So Fine?....................... 182
Beaches - Here Today, Gone Tomorrow............... 185
Beaches - Moving On................................ 188
Moving Mountains of Sand - Dredging
    California's Harbors............................ 191
Dams - Cutting off our Beach Sand.................. 194
Messing Around with Beaches....................... 197
Beach Sand Burglary................................ 200
Gold in Beach Sand................................. 203
The Colors of Beaches.............................. 206
Moving Sand Around................................ 209
Italian Beaches..................................... 213

Part V: Waves, Currents & Sea-Level Rise
Sunlight and Sea Level - What's the Connection?..... 218
Keeping Track of Sea Levels......................... 220
Return to Pleasure Point - Again.................... 223
Giant Waves at Sea................................. 226
Rogue Waves....................................... 229
Lost Cargo Tracks Ocean Currents................... 232
Floating Frisbees and Coastal Currents.............. 236
Aerial (mis)Adventures............................. 240
Coastal Currents and Missing Sailors................ 244
Sea-Level Rise - Should We Worry?.................. 248
Sea-Level Rise - Sorting Out the Pieces.............. 252

Predicting Future Sea Level 257
Big Wave Surfing 262
Are Waves Getting Bigger? 265
Sea-Level Rise-Searching for the Answers 268
Sea-Level Rise-What Can We Expect? 271

Part VI: Natural Disasters

Tsunamis-Should We Worry? 276
Subduction Zones, Great Earthquakes, and Tsunamis .. 278
Cascadia-A Sleeping Giant 281
Tsunamis-To Worry or Not to Worry? 284
Lituya Bay-The Largest Wave Ever Witnessed 287
Holding Back the Sea 290
Superstorms and Storm Surges 293
Weather on Steroids 297
Perspectives on Disaster 301

Part VII: Marine Life

Algal Blooms-Good and Bad 306
Global Migrations-Sooty Shearwaters 309
Tracking Marine Mammals 311
High-Tech Ocean Observations 314
Reading a Salmon's Ear Bone 316
Calamari-Small, Large, and Extra Large 319
Ocean Economics 322
Sharks-Who's Attacking Whom? 325
The Biggest Animals That Ever Lived 329
The Life of a Blue Whale 333
Fertility, Food Chains, and Fish 337
Crabs and Calamari 340
Invasion from the Sea 343
Aquatic Aliens 346
Peregrines, Pelicans, and Pesticides 350
Pesticides and Peregrines-Making the Connection 352

Bringing Back the Peregrines
   from the Edge of Extinction........................ 355
Offshore Feeding Frenzies................................ 358

Part VIII: The Global Ocean

Heading Across the Pacific............................... 362
Crossing the Pacific Plate................................ 365
Volcanoes in and Around the Pacific................. 368
Perils of Foreign Ports..................................... 371
Traffic at Sea... 374
Containers at Sea... 377
The Birth of the Indian Ocean......................... 380
Stories from Ancient Rocks.............................. 383
Finding Yourself at Sea.................................... 386
Iceland - The Making of an Island..................... 389

Part IX: Ocean Exploration

Drilling into the Sea Floor............................... 394
Ocean Drilling - Asteroids, Mass Extinctions
   & the Mediterranean.................................. 397
Ocean Drilling Confirms Continental Drift......... 400
Discovering the Seymour Center...................... 403
Venturing into the Arctic................................. 406
Nansen's First Arctic Adventure....................... 410
Nansen Heads for the Arctic............................ 414
Drifting Towards the North Pole...................... 418
Nansen's Dash to the Pole............................... 421
Trying to Find Franz Josef Land....................... 424
Nansen Trying to Get Home............................ 426

Part X: Climate Change

Climate Oscillations and Disappearing Sardines....... 430
What's Next For Our Shoreline? El Niño and La Niña... 433
Calera - Green Cement for a Blue Planet................. 437

Climate Change and Rising Sea Level 439
Messing with the Atmosphere 441
Greenhouse Gases and Climate Change 443
Collapsing Cliffs and Shifting Climate 445
Santa Cruz - Going to Extremes 448
Changing Climates and Shifting Species 451
Weather, Climate, and Coastlines 454
A Melting Arctic 457
The Hazards of Arctic Drilling 461

Part XI: Energy & Power

Oil Formation in the Sea 466
Inside the Ocean's Oil - Making Machine 468
Energy from the Oceans 471
Energy and the Oceans 476
Energy from the Waves 481
Getting Real About Wave Energy 485
What's Up with Tidal Power? 489
Ocean Energy from Warm Water and Wind? 492
A Drilling Disaster 496
Drilling and Spilling 499
Shuttering San Onofre 501
A Nuclear Power Plant for Santa Cruz 504

Part XII: Water Fresh & Salty

Why the Oceans are Salty 508
Five Years and 130 Columns 511
A Primitive Ocean 513
Salt in the Sea 515
Salt from the Sea - Part 1 517
Salt from the Sea - Part 2 520
Fresh Water - A Drop in the Bucket 523
Rainfall, Runoff, and Reservoirs 526
Groundwater - An Invisible Resource 529

Groundwater - Out of Sight but Not Out of Mind 532
Droughts and Mega - Droughts 535
Reading Tree Rings 537
Water - Searching for Answers 540

Acknowledgements

Any book or any collections of essays or stories, by its nature, involves many different experiences and people, in this case, stretched out over the 46 years of my life here on the edge of Monterey Bay. To thank all of those who have provided ideas, insights or information from their own lives and adventures would tax my memory greatly, and would ultimately leave out many I can't recall. So I have chosen to thank those whose time and input have been the most important in bringing these bi-weekly newspaper columns together into this book, and more broadly thank all of the others - family, friends, colleagues and those who have given me topics, questions and mysteries to solve.

One person, in particular, I want to thank for her dedication to making this book possible. Deepika Shrestha Ross, my partner, encouraged me to put these 170 different essays, written over six and a half years, into a book form so readers would have all the stories together in a single place. The book wouldn't have happened without her boundless patience and considerable design skills. For this I am deeply grateful.

While I have covered bits of history in many of these essays, I have tried not to tread too hard on the feet of my old friend, frequent traveling companion and local historian, Sandy Lydon. Whenever history was involved, I would usually check with Sandy first for a short answer to a question I wanted to get right. More often than not, Sandy took the time to respond with a (lengthy) short course in local history on the particular topic. Thank you Sandy, for all of your input. May you continue to enrich the local community for decades to come.

In 2012, Kenneth and Gabrielle Adelman decided to document the entire 1,100-mile California coast in high-resolution digital photography taken from their own helicopter. They posted these images on a rather amazing website (www.CaliforniaCoastline.org), which has been extremely useful in my work over the years and the work of others as well. In subsequent years they have reflown the coast in seven subsequent years and added some older photographs as well. I am appreciative of their dedication to this project and I have used a number of their images in this collection of stories.

Importantly, I want to thank the Santa Cruz Sentinel, in particular, Don Miller, Julie Copeland, Conan Knoll, and Don Fukui, who have continued to publish these columns over the years. I also want to acknowledge and thank my partner in Our Ocean Backyard, Dan Haifley, who has been writing an article every other week all these years. And not to be left out, thanks to all of those readers, friends and others, who have responded to columns, posed questions and suggested new topics to consider. You have all been important and these stories are not finished yet.

Gary Griggs
November 2014

Foreword by Sam Farr

The book you hold in your hand is not just a local story - it is a national story. It tells the story of America's coastline and America's ocean. The California coastline is central to both our quality of life and economic well-being as Americans because of its abundant resources - rich marine life, four prized National Marine Sanctuaries, and the busiest container ports in the nation. People are drawn to this spectacular coastline to either build their lives and their fortunes, or to visit and explore. This is why California has, by far, the largest coastal economy of any state and why, back in Washington, I like to remind folks that protecting the environment makes good economic sense.

But what happens in Washington starts in the district. It starts with raising public awareness. It starts with the kind of stories that Dr. Gary Griggs laid down on these pages. Gary is a talented interpreter of complicated coastal science and has a long track record of effectively translating his findings for policymakers. Gary's biweekly column for the Santa Cruz Sentinel has been instrumental in promoting popular understanding of these important issues across the Monterey Bay region and beyond. We need to better understand our environment in order to fully recognize its value and protect it. That is why I am so pleased to see Gary's essays collected and published here, to be shared with the regular readers from our Santa Cruz community and a broader audience of new readers everywhere.

A leader and a visionary in our community for more than four decades, Gary has actively sought to break down silos and bring people together. The UC Santa Cruz Long Marine Lab campus at the edge of Monterey Bay physically embodies this collaborative approach. It houses the Seymour Marine Discovery Center, the Center for Ocean Health, and the Predatory Bird Research Group. The lab shares space with federal, state, and private entities, including the National Oceanic and Atmospheric Administration's Southwest Fisheries Science Center, the State of California's Oiled Wildlife Veterinary Care and Research Center, and Island Conservation. This ever-growing center of expertise is testimony to Gary's visionary leadership and his ability to translate the story of our treasured coastline into concrete action to study and protect it.

Gary's words and leadership make the ocean relevant for Californians and Americans alike. This collection of essays is evidence to that and offers you the same opportunity I have had to be inspired by both Dr. Gary Griggs and the spectacular coastline he explores.

Sam Farr
Representative
October, 2014

Foreword by John Laird

In California, the hot spot for biological diversity in the United States, we have many natural resources that are unique and special. Our long and winding coastline - along with the ocean that borders our state - is one of those natural jewels. It is the heart of a unique climate, vibrant ecosystem, and our state's economic life. It is the reason many of us choose to live by the coast.

There are many threats to the ocean: acidification caused by climate change, the growing amount of marine debris, historic fisheries challenged by overfishing and environmental changes, and the impact to coastal residents and infrastructure from sea-level rise. Yet one of the biggest threats is a lack of understanding of these challenges - how they came about, what we can do to change them, how we can motivate others to work with us to do just that.

One of the pleasures of being California's Secretary for Natural Resources is taking leadership in our state's ocean policy. The challenges are huge. In trying to meet them, scientists who can explain issues clearly, talk from their own experience, and form the basis for an informed public to take action, are jewels in their own right. Gary Griggs is special in this regard.

As a Santa Cruz newspaper reader, I read many of the columns that appear in this book as they were published. I marvel at Gary's ability to relate more complex situations in an understandable form for lay people - and in a way that holds their interest. It is obvious why he has been a successful and much-loved teacher for over four decades. It is a great thing that many of his columns now appear in one place, available to an even wider audience.

I hope you will enjoy these columns as much as I have - and I hope it will motivate you to take action to make sure we turn over an ocean that is environmentally sustainable to the next generations - in a way that continues to give the joy and the sustenance that it has given to all of us.

John Laird
California Secretary for Natural Resources
October, 2014

Preface

OUR OCEAN BACKYARD

The First Six Years

Six and a half years ago I started writing a bi-weekly column, Our Ocean Backyard, for the Santa Cruz Sentinel. Santa Cruz sits smack on the ocean and many of the people who have chosen to live here have done so because of Monterey Bay and the much larger Pacific Ocean beyond. In a state that in many ways is defined by its coast, it isn't surprising that so many people are drawn towards our ocean edge. Whether it is our weather and climate, or any number of recreational pursuits: fishing, surfing, sailing, jogging or walking, or just observing and exploring, the ocean provides something for all of us. Each of these different pursuits often generates questions, as does simply living in a community facing the sea.

When I was first asked to write the column I felt that I could probably find enough things to explore for about six months, maybe. But, over six years later, the questions and topics seem to keep coming- often from newspaper headlines, occasionally from National Public Radio stories, or from readers or friends who suggest something that they have always been curious about. The list of potentially interesting ocean backyard topics, whether local, regional or global, is seemingly endless: natural and human history, ocean explorers and early residents, environmental disasters, marine life, changing beaches, coastal landforms, climate change and sea-level rise.

There are really two major reasons or motivations for the Ocean Backyard columns and this compilation. By the way, neither includes getting rich or famous, although there is always that hope. On the one hand, I believe almost everyone of any age has some natural curiosity about the world around them. And I hope with these columns that at least occasionally, I touch on a topic or subject that you have wondered about. And while most of the columns are by newspaper demands quite short, it is my goal to try to explain each topic in a way that all readers get a clearer sense of why things are like they are.

The second reason is what many perceive is the need for better science communication. If you haven't gathered from reading the columns over the past six years, I am a scientist, actually a coastal geologist, and so it is from a scientific perspective that I approach each story. Interesting and surprising to me, while the "average" American (I am never sure just who this refers to), puts much more faith in or holds scientists in much higher regard than say, politicians, journalists, or the clergy, surveys repeatedly reveal that this same average American doesn't really understand much about science.

A 2006 survey by Research!America, an organization that supports science research, found that 60% of Americans say that a scientist is an occupation of "very great" prestige, the highest of any job category surveyed. I have to say I was very happy to read this conclusion. The comparable figure for doctors is 58%, firemen 54%, members of Congress 31% and journalists 16%. This survey was done in 2006, however, so prestige of members of Congress may have slipped a bit.

Corey Dean, the former science editor at the New York Times has written an useful book entitled "Am I Making Myself Clear? A scientist's guide to talking to the public". She points out that in spite of the respect that the public holds for scientists, in general, most Americans lack an understanding of scientific or technical information, which is why so many embrace astrology, UFOs, and creationism. Less than half of all Americans (about 45%) accept the theory of evolution, yet most believe that astrology is at least somewhat scientific and a majority believes that some people possess psychic powers. Evidence for this is that virtually every newspaper publishes a daily horoscope, while few devote any regular coverage to science.

Two other recent books also speak to the problem that we as scientists don't do a very good job of communicating what we do to the public in an understandable and interesting way. Randy Olson left behind his professor position in marine biology to go to film school and now works on communicating science. The title of his book - "Don't be such a scientist"- pretty much says it all.

Nancy Baron is the lead communications trainer for the Packard Foundation funded COMPASS program (Communication Partnerships for Science and the Sea) and has helped thousands of scientists learn to communicate more effectively with the media, decision makers and the public. Her book "Escape from the Ivory Tower- A Guide to Making Your Science Matter", came from this experience.

Obviously, we as scientists need to do a much better job and take a far more active role in explaining what we do and why it's important. While I personally believe all scientists should be able to explain what they do and why this is important in a 30-second elevator speech, this is far easier for someone studying the ocean than someone who studies superstring theory or bioinorganic chemistry.

I have continued to write these stories with the belief that they are making what happens along our coastlines, in our oceans, and on the adjacent land more understandable to those who read them. Some readers who only just discovered, or began reading the columns recently, often ask: "Why don't you write about beaches, or submarine canyons?" And it often turns out that I had covered these topics several years earlier.

So I thought it would be useful to go back over all the articles, updating and editing them where changes have occurred, adding photographs or graphics where it helps to explain the story, before assembling them between two covers. And whether you are well into the age of e-books, or still enjoy holding a traditional hardcopy, this collection of the first six and a half years of columns are available both ways. I hope these stories will reinforce and enrich your own sense of place, living, as we so fortunately do, with Monterey Bay and the Pacific Ocean as "our backyard."

PART I

HISTORY
OF THE
MONTEREY BAY REGION

Article 1

Exploring the North Coast

Wood trestle spanning San Vicente Creek, Davenport, c. 1906. (courtesy: Sandy Lydon)

Driving north from Santa Cruz up Highway 1 is almost always an interesting experience for me, perhaps because I have this thing about the coast. There is usually something new or unexpected, even though I've driven that road hundreds of times. The early Spanish explorers, however, didn't see the journey up the coast quite the same way. If fact it was the journey from hell by their written accounts.

Today Highway 1 is two or sometimes three, very smooth, nearly level lanes, and you can easily drive 50 or 60 miles per hour, unless, as is often the case on the weekend, there is someone soaking in the scenery and driving very slowly in front of you. What you don't usually notice from your car is that each of the many stream valleys or canyons that emerge from the coastal hills is crossed by an embankment or fill that the highway passes over. There are a number of these stream valleys that you just zip right across without giving them a second thought, Wilder Creek, Majors Creek, Laguna Creek, Yellow Bank Creek, San Vicente Creek, to name a just a few of the larger ones.

But 245 years ago, when Captain Gaspar de Portolá and his tired and scurvy-ridden traveling companions started up the north coast on horseback from what was later to become Santa Cruz, the route across each of those steep, brush covered stream canyons was just about impassable. As they struggled along the route followed by Highway 1 today, they had to work their way down one side of each steep canyon, through poison oak, willows, and oak trees, across the valley floor, and then drag their horses and mules up the opposite side. They weren't happy campers and I'm sure that at least a few of them experienced the added discomfort of poison oak and ticks.

The Portolá expedition had been sent north from San Diego, overland, to find the glorious harbor named Monterey reported by Sebastian Vizcaíno from a ship 165 years earlier. While Portolá and his men looked out over the bay from Mulligan Hill at the mouth of the Salinas River in 1769, they didn't see anything through the fog that looked remotely like a harbor. This is why they found themselves bushwhacking their way up the north coast, searching for an anchorage for Spanish galleons returning to Acapulco from Manila. These are some of the stories that Sandy Lydon brings to life for his local history classes and field excursions. Through our local adventures together, some of this recent history has actually started to rub off on me, although he cautions me about practicing history without a license.

In 1905, a bold, but ultimately unsuccessful effort was initiated to ease the trip north or south along the coast. The Ocean Shore Railroad was begun as a way to connect Santa Cruz to San

Francisco along the coast and lay the way for developing a number of planned communities along the San Mateo County coast. Excavation, filling and then laying a double set of tracks began at both ends, but was met with some overwhelming challenges in the middle.

For starters, there was a formidable task of building a rail line across Devil's Slide. Somewhat surprisingly, the engineers of the day managed to actually build tracks across the slide, but the great 1906 San Francisco earthquake carried much of that portion of track into the ocean, which was a huge setback. Construction continued at the northern end, however, and by 1908 the railroad was running passengers, and freight, including artichokes, hay, horse beans, potatoes, dried peas and canned cabbage. Oil pumped out of the ground near Half Moon Bay was also shipped out in barrels on the Ocean Shore Railroad.

At the Santa Cruz end of the line, the firm of Shattuck and Desmond was contracted to build a line from Santa Cruz to Scott Creek, where the two lines were to connect. The completion of the double-tracked, electrified Ocean Shore Railroad was to be announced in early 1907. The plans for crossing all of those stream valleys that the Portolá expedition struggled through in 1769 involved initially building wooden trestles. The trestles were used by the train as a way to transport rock and soil taken from the cuts to build a huge fill embankment across each stream valley, which would then support the weight of the rail line. In addition to Devil's slide at the north end, Waddell Bluffs plunged directly into the ocean a few miles north of Scott Creek and became an insurmountable obstacle for the ill-fated Ocean Shore Railroad.

The rest of the story will have to wait.

Article 2

Forty Miles of Bad Road

Construction of Highway 1 across Waddell Bluffs, Santa Cruz County, c. 1945. (courtesy: California Department of Highways)

The North Coast article two weeks ago generated a lot of comments and questions, and there are a lot more stories about the northern part of Santa Cruz County; asphaltic sandstone and petroleum, stage coaches and Stanley Steamers, railways and highways, to mention a few.

It wasn't until 1947 when Highway 1 was built across the base of Waddell Bluffs that Santa Cruz was first permanently connected to San Mateo County to the north. Thirty square miles of the southern tip of present-day San Mateo County, including Año Nuevo Point,

Pigeon Point and Pescadero, were originally part of Santa Cruz County. But because the barrier at Waddell often prevented access to the county seat in Santa Cruz for those northerly homeowners, this area was annexed by San Mateo County in 1868, although not without a fight. At that time you couldn't always get here from there.

Sandy Lydon sent me a description of the road between Santa Cruz and Pescadero from an 1865 Santa Cruz County newspaper: "....it is a great hardship and injustice to the people living in Pescadero and its vicinity in compelling them to go to Santa Cruz upon all county business, a distance of 40 miles over one of the most abominable roads this side of Kamchatka - a road, in fact, that is totally impassable at times, either on foot or on horseback, as a portion of the distance must be traveled along the beach, which is encompassed by a high bluff upon one side and the foaming billows upon the other, and which is completely covered by surf in every southern gale…"

The first regular stage coach connection to the north began in the early 1870s, and for years made the trip daily, usually pulled by two horses, sometimes four. The stage went north from Santa Cruz to Pescadero, and then headed inland through the mountains to San Mateo and Redwood City. It carried everything from shovels and sacks of flour to needles and thread for coastal housewives, and even small batches of cheese and butter from the old Steele Dairy north of the county line destined for markets in San Francisco.

Passengers and the stage often had to wait for hours at Waddell Bluffs for the tide to drop. Even then, during large waves, the water would wash up around the horse's feet. Adding to the excitement of early travel along the coast were the rocks that constantly fell and rolled off the bluffs and onto the beach.

The road along the coast was just a dirt wagon track in those days, and was either a long dusty ordeal in the summer months or a muddy trek in the winter. Although the Ocean Shore Railway was planned to shorten the trip and transport more visitors, the sections built at both ends were never connected. Waddell Bluffs remained a formidable barrier and the tracks heading south ended at Tunitas Creek. Passengers coming from San Francisco were loaded onto the stage, which still had to navigate across the shoreline at low tide.

The tracks heading north from Santa Cruz extended 15 miles to Swanton where the railroad curved inland in an abortive attempt to avoid what was then called Gianone Hill. With the coming of the automobile, the stagecoach connecting Tunitas and Swanton was replaced by a Stanley Steamer bus.

In 1905, engineers with the Ocean Shore Railway experimented with some rock-filled timber cribbing at the base of the Waddell Bluffs to see if they could find a way to resist wave action, but this also proved unsuccessful. There was so much rock constantly coming off the bluffs that it was one huge talus pile from just beyond Waddell Creek to the county line. Wagon roads were built but never lasted for long between the waves battering them on one side and the bluff failures on the other.

It was 178 years from the time the Portolá expedition clawed their way up the coast in 1769 until the California Department of Highways was able to build a permanent road across the base of the bluffs. About a million cubic yards of loose rock (or about 100,000 dump truck loads) were removed from the base of the bluff and 600 feet of riprap was placed to protect the area most exposed to wave attack. Much of that rock survived until the 1983 El Niño winter. Large storm waves combined with elevated sea levels and high tides removed the loose rock and fill protecting about 2000 feet of highway at the south end of the bluffs. Cal-Trans brought in twenty-four thousand tons of rock to save Highway 1 that year.

ARTICLE 3

The Old Coast Road

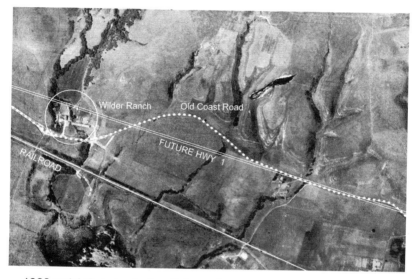

1928 aerial photograph of north Santa Cruz County highlighting Highway 1, the Old Coast Road, the railroad, and Wilder Ranch.
(© D Shrestha Ross, photo: former Ocean Shore RR, Gary Griggs Collection)

While the Santa Cruz section of the old Ocean Shore Railroad was pretty much a straight shot from Nearys Lagoon to Swanton, the Old Coast Road was a different story. The railroad required long straight stretches that were nearly flat or had very low grades, which meant it had to get across each of the stream valleys at the same elevation as the terrace on either side. This was achieved by first building timber trestles spanning each valley, and then filling around these to create massive embankments that essentially dammed all of the streams.

This did require some engineering, as well as a solution for getting the now dammed creeks under the embankments to the ocean. Somewhat puzzling was the approach used back in 1905. For some reason, not completely clear to me, they chose to dig tunnels through the mudstone bedrock on the north side of each stream valley, rather than build large culverts. Maybe they hadn't invented culverts yet; but if necessity is the mother of invention, the work of hand digging a tunnel through solid rock for a creek to flow through on its way to the ocean might have been an incentive to design the first culvert. But the tunnels work and remain as interesting landmarks of the coming of the railroad and the rerouting of the original streams a century ago. You can easily find them emptying from their bedrock tunnels onto the beach at the north end and ocean side of every one of those embankments. You can even walk through them, if you are adventurous and have a rubber boots and a good flashlight. If there were any steelhead or salmon that originally spawned in these creeks, the embankments and tunnels unfortunately terminated their migratory activities.

The road up the coast in the first half of the last century didn't really go anywhere because it was ended at Waddell Bluffs, so traffic and use were limited. It started as a wagon road, was gradually paved, and took the easiest route in order to minimize bridges or fills. The original road can be clearly seen on the old aerial photographs (and the oldest aerial photographs of the county, taken in 1928, are now on-line on the California Coastal Records Project (http://www.californiacoastline.org/).

Mission Street extension marks the beginning of the Old Coast Road. It headed north from Swift Street past the packing sheds and the old Wrigley building to Moore Creek, where it angled uphill and crossed over the creek. It continued inland a ways and then turned north and essentially followed the present Highway 1 path for about half a mile. At this point, just past the horse stables, there is a driveway that turns off on the right side. This was the old highway, which then swung north and ran along the base of the hill. The old road then turned towards the coast and crossed the present Highway 1 right where the bike path enters Wilder

Ranch today. As you pass around the gate and start down towards the old ranch buildings you are on the Old Coast Road with the white center line still visible. You can follow the original route as it crosses Wilder Creek and then passes right in front of the old ranch houses. The road heads back up the hill to join Highway 1 where vehicles enter the state park today. It then turned north to follow the present highway alignment.

For the next several miles, until the little settlement of Majors, the Old Coast Road followed the route now occupied by Highway 1. It crossed Majors Creek where it does today, on a wide earth embankment because it was constrained by the railroad on the ocean side and the steep canyon walls just inland. After crossing the creek it followed a route closer to the railroad; one that you can still drive on today. The old road passed through Majors, which was a former stop on the Santa Cruz to Davenport line of the Southern Pacific Railroad. This area was earlier known as Enright, named after Joseph Enright who owned a thousand acre ranch and dairy here. He was a member of the County Board of Supervisors and also an organizer of the Santa Cruz Oil Company. The old road intersects the present highway at a very high steep cut that was part of a highway straightening effort in the 1940s and 1950s.

At each of the next four creeks, Laguna, Yellow Bank, Liddell and San Vicente, the old road turned inland so as to avoid the problem and cost of having to build a huge embankment across each stream valley. Instead the Old Coast Road descended down the side of each valley and crossed over each creek further upstream on small bridges. The old road and bridge still exists at Laguna Creek where it is still called the Coast Road. Time has eliminated parts of the original surface from the Yellow Bank and Liddell crossings but each of these old curved inland remnants are still clearly visible on Google Earth.

Article 4

Hydrocarbons in the Hills

City Street Improvement Company asphalt mine, Santa Cruz, 1914.
(courtesy: Santa Cruz Museum of Art & History)

It's the climate, coastline and beaches that drew many of us to the Monterey Bay area and that has drawn visitors for well over a century. Main Beach and the Boardwalk, surfing and sailing, fishing and fish restaurants, and the rest of the visitor serving establishments are at the core of our local economy.

In the late 1800's the economy was based on some other natural resources; instead of sand, sun and salt water, it was lime, leather and lumber. The production of cement from the marble on both the Cowell Ranch and in the hills above Davenport was a big part of the Santa Cruz County economy for a century. Another geologic resource from the north coast, less well known and with a shorter history, was hydrocarbons - natural asphalt and oil.

The first paved roads built in the world were of rock asphalt. Paris reportedly paved its first street with this material in 1854. Not to be outdone, in 1872 Union Square in New York City became the first street paved in the United States with rock asphalt brought all the way from Switzerland. Two decades later in 1891, Salt Lake City paved its first street with asphalt brought from Santa Cruz. The 1890s also saw asphalt from the Santa Cruz quarries shipped to San Francisco, and also Seattle, to pave streets.

You might ask, asphalt in Santa Cruz? A State Mining Bureau report from 1894-96 describes asphalt beds located on Rancho Refugio between Majors and Baldwin creeks. "About six miles northwest of Santa Cruz and near the ocean are extensive beds of bituminous rock, which at present are being successfully quarried and utilized for making pavement. In fact, it is regarded by those in a position to know best, that bitumen is the pavement of the future, and in the beautiful city of Santa Cruz it is the pavement of the present. These beds of bituminous rocks cover an area perhaps a mile square, and are the residuum of oil beds in a period not geologically remote...It is in the head of a canon on the Baldwin Ranch, 5 miles northwest from Santa Cruz and 2 miles from the coast at 500 feet elevation...A good road has been made from the mine to the county road on the coast." This area was known at various times as the Asphalt Beds, the Petroleum Works and later the Baldwin Mine.

Today driving north on Highway 1, you can glance up Majors Creek Canyon, just before you reach the former red, white and blue mailbox on the opposite side of the road. Looking inland and up that canyon you can see massive, steep black cliffs, which are composed of bitumen-saturated sandstone, or rock asphalt. A large portion of the cliffs on the north side of the canyon collapsed in a massive landslide in 1960, damming Majors Creek and forming a small lake. A few hundred yards further, a turnoff on the left takes you through the little community of Majors, which used to have a small school known as the Petroleum School.

The rock asphalt or bitumen-saturated sandstone in this area was injected from below into the Santa Cruz Mudstone in a fluid state. Numerous sandstone dikes and sills, many of which contain bituminous material, are exposed in the seacliffs and road cuts between Wilder Creek and Greyhound Rock. The Santa Margarita Sandstone, thought to be the source of these intrusions and a geologic formation quarried for sand across Highway 1 from the entrance to Wilder Ranch State Park, contains varying amounts of bitumen. The hydrocarbons are believed to have migrated into the Santa Margarita Sandstone from the underlying Monterey Shale, one of California's largest oil producers.

The asphaltic content of the sand ranges from about 4 percent to as much as 18 percent by weight. These oil-impregnated layers are up to 40 feet thick and when sufficiently warmed by the sun, tar may drip out of the bituminous sands. An estimated 614,000 tons of asphaltic paving material, worth approximately $2,360,000, was produced from this area between 1888 and 1914. Production was intermittent after the 1920's, with the last of the quarries ceasing operations in the 1940's.

The presence of the asphalt along the north coast, and the discovery of oil in southern California led to the drilling of a number of wells on the north coast terraces in 1901. While fortunes were expected, the amounts recovered were small. Local consulting geologist and old friend Jerry Weber reports that in 1955, Husky Oil Company in partnership with the Swedish Oil Shale Company, began an experimental project using gas-fired burners inserted into shallow wells in the asphalt sands in an effort to liquefy and extract petroleum. Over a 3-year period, a total of 228 burner-producer wells were drilled in these hills, raising down-hole temperatures to 600 degrees Fahrenheit. When completed, 2,665 barrels of oil and 4,520 million cubic feet of natural gas were recovered. While this was a reasonable recovery rate, fuels costs and high heat losses made this an uneconomical project.

Article 5

Ships of Cement

The SS SantaCruzCement
at the end of the 2,600 foot pier at Davenport, c. 1940s.
(courtesy: Alverda Orlando)

Santa Cruz County is probably one of the only places around that has had three large "cement" ships in its coastal history, none of which was made solely of cement. All of the ships dated from the early part of the last century, and each was used for different purposes.

One of the natural resources of the Santa Cruz Mountains that was exploited early on was limestone, actually marble, a basic raw material for making cement. When the Davenport cement plant was being constructed in 1905, there wasn't yet a railroad up the north coast, but it wasn't long until trestles were built across the canyons and tracks were laid. A direct rail connection up the coast to San

Francisco was never completed, so cement had to be transported through a longer and more indirect route, south past Watsonville, through Chittenden Pass, and then north to the bay area.

An idea was hatched, which seemed like a great solution at the time - transport the cement by ship directly to Stockton where it was to be offloaded. So as only engineers can, a pier was boldly designed and constructed in the early 1930s starting from the sea cliff opposite the cement plant. A pile driver was lowered down the cliff and steel piles were driven across the beach and then offshore into the seafloor. These were then encased in concrete to resist attack by salt water. Ultimately a 2,600-foot long pier was completed, although wave damage during construction led the engineers to turn the outer end of the pier slightly more northerly into the swell, hoping to reduce the wave impact. The pier had two 12-inch diameter pipes for loading dry cement as well as a 3-inch line for water and a 6-inch pipe for fuel oil. Because the pier intersected the coastline part way down the sea cliff, a tunnel was constructed completely through the cliff to connect with the silos where the dry cement was stored on the opposite side of the highway.

The SantaCruzCement was a ship specially outfitted to carry cement, and would be tied up to a series of buoys just offshore from the end of the pier. Cement was loaded through a flexible hose from the end of the pier as wind and wave conditions were nearly always too rough for direct tie up to the pier. The ship was in service for a little over 15 years until wave conditions were deemed too dangerous to ship cement safely by sea. Wave impact over the subsequent decades has progressively eaten all of the piles but the four sets, which are still visible from the cliff top, as are the large steel doors in the sea cliff at the mouth of the tunnel.

About 18 miles southeast of Davenport lies the county's second "cement ship". To be honest, however, it should be called the concrete ship as it was constructed of concrete, reportedly from Davenport, rather than just cement. This vessel, the SS Palo Alto, was designed as an oil tanker and was built to be part of a fleet of concrete ships built in 1918-1919 for the war effort. World War I ended in 1918 before the ship was completed, however, and ultimately it was sold as surplus to the Seacliff Amusement Company of Nevada and towed from San Francisco to Seacliff Beach in 1930.

It was sunk in shallow water with its bow facing into the sea, and a 600-foot long pier was built to connect its stern with the shore. For the next two years, it was an amusement center and party boat, with a ballroom, restaurant, concessions stands, and even a swimming pool. The large waves of the winter of 1932 broke the ship's back, however, which led to closing it down and the beginning of its economic decline. The ship's mechanical equipment and superstructure were sold a few years later to a local wrecker for scrap and salvage. In 1936 the State of California in a bargain sale bought the ship's hulk for one dollar and incorporated it into Seacliff State Beach, where it became a favorite fishing platform for decades.

Storms ultimately cracked the ship further, and due to continuing deterioration, the ship itself was finally closed. In the spring of 2005, 75 years after being sunk in shallow water, oil found on wildlife in the area was traced back to the ship. Clean up operations were initiated. No oil was known to have spilled but birds were believed to have come into contact with the oil by entering the ship's cracked hull while diving underwater.

A little known fact that my friend Sandy Lydon shared with me recently, the SS Palo Alto had a concrete sister ship, the SS Peralta, also built as an oil tanker for the war effort, but that never saw military service either. It was converted to a fish cannery, used in Alaska for a while, but also anchored at various points around Monterey Bay buying fish and canning them right on board in 1926. This caused grief for the local canneries in Monterey and eventually ended up in court. In 1958 the SS Peralta was purchased by Pacifica Papers to be part of a giant floating breakwater built to protect a pulp and paper mill on the Powell River in British

Columbia. The ship remains there today, and at 420 feet long is believed to be the last of the World War I fleet and also the largest concrete ship still afloat.

Article 6

Shipwrecks in the Sanctuary

The La Feliz, forced onto the rocks
just north of Natural Bridges State Beach, Santa Cruz, 1924.
(courtesy: Santa Cruz Museum of Natural History)

On the night of October 1, 1924, the combination of high seas and a course too close to the shoreline put the La Feliz on the rocks directly in front of where Long Marine Laboratory exists today. The 100-ton vessel was carrying canned sardines from Monterey to San Francisco when she was wrecked. Local residents drove out to the top of the 30-foot bluff and used their headlights to illuminate the ship and help rescue the crew of 13. The mast was removed, leaned against the cliff and used with a block and tackle to recover the cargo of sardines as well as equipment from the ship. Somewhat surprisingly, comparing photographs of the shipwreck with the site today, neither the rocky shelf where the ship was grounded nor the cliff has changed much in the subsequent 90 years.

Standing today on the deck of the Seymour Marine Discovery Center at Long Marine Laboratory, affectionately know as Shipwreck Deck, you can see what looks like a tilted telephone pole, rising from the shoreline and extending up above the cliff top. Amazingly, this is the mast of the La Feliz, still standing proudly, 90 years later. While 2000 feet away at Natural Bridges, two of the three arches have collapsed over the years; the cliff in front of the Marine Discovery Center remains intact. One important reason for this difference is the presence of a very hard rock platform in the Santa Cruz Mudstone at the base of the cliff. The very resistant rock that impaled the La Feliz has protected its mast and also buffered the adjacent cliffs from direct wave attack. On low tides you can often see the remains of the ship's drive shaft on the beach just east of the mast.

The La Feliz wasn't the only local shipwreck. Forty-eight years earlier in October 1876, the Active, a 92-foot schooner, went aground on Its Beach just below the old lighthouse. The Active had taken on a load of railroad ties from the wharf in the morning but didn't set sail until 8 pm that evening. Shortly after getting underway, however, the wind died, and left her about a mile off the lighthouse. When the wind came up later in the evening, and as the crew was working to get underway, the Active was hit suddenly by heavy seas. Even dragging two anchors, the waves still washed the schooner onto the beach. At daylight a line was thrown from the cliff below the lighthouse to the vessel. The crew used the line to work their way to shore, hand over hand.

October seems to have been a bad month for shipwrecks. On October 11, 1912, a submarine named F-1 was washed ashore at Port Watsonville, which was a short-lived wharf located at the end of Beach Road. Two seamen were washed overboard and drowned.

Looking at a statewide database of shipwrecks reveals that the Santa Cruz County coastline has been a whole lot safer for ships than many other sections of coast. Santa Cruz comes up pretty infrequently compared to Humboldt, Mendocino, San Francisco, the Channel Islands and the Monterey Peninsula. A catalog of ship losses completed by the Monterey Bay National Marine Sanctuary lists 463 individual vessels, but only 25 of these met their end along the coastline of Santa Cruz County. Most of these were listed as "foundering" or "sinking offshore", with a handful of others stranded on the shoreline and two burned at sea. The SS Palo Alto at Seacliff was the only vessel that was intentionally stranded and also the only one still intact today.

The Active, grounded on Lighthouse Beach, Santa Cruz, c. 1876.
(courtesy: Santa Cruz Museum of Natural History)

Article 7

California Whaling

Inflated humpback whale
being pulled ashore for processing, San Francisco Bay, c. 1900.
(courtesy: Edie Rittenhouse)

One of the high points each year when I was growing up in southern California was heading north in the old family station wagon for a month of camping every summer in the redwoods, and along the Oregon and Washington coasts. Our first stop was always in Berkeley where my dad's old college roommate and his family lived. He was on the faculty there in Geography for forty years.

Jim Parsons was a fascinating guy who took his family all over the world on his research trips. They had a great big house in the Berkeley hills filled with books and all sorts of interesting things he had collected from various adventures to exotic places around the world. Each summer when we visited them I thought seriously about one day being a college professor. I knew he had a doctor's degree, which at age 10 was a little intimidating for me to contemplate. So I asked my dad, "What do you have to do to get a doctor's degree?" To which my dad replied, "You have to write a book!" Not being much of a writer at age 10, I quickly shelved the idea of ever becoming a college professor.

After an always interesting evening, we would leave Berkeley early the next morning, and head north along the edge of San Francisco Bay to Richmond, where we drove the car onto a ferry for the trip across to San Rafael and the beginning of the Redwood Highway. This was in the early 1950s, well before the Richmond-San Rafael bridge was built. Driving the car onto a ferryboat with the sound and smell of the diesel engines, the boat rolling with the swells, the whistle as we cast off, and then standing on the upper deck for the bay crossing was always pretty exciting.

On one of those early trips, I clearly recall my dad stopping the car along the waterfront area of Richmond and saying- "Hey, I want to show you guys something". My two brothers and I were always up for getting out of the back seat of the car, so we jumped out and followed my dad across the highway. I have a vivid memory of an overpowering smell, and then the sight of a massive whale up on a ramp. I was probably 10 or 11 at the time, yet the sight and smell that morning is still pretty vivid in my mind.

We were seeing the Richmond whaling station, one of the last two whaling stations in America, which was active in San Francisco Bay until 1971. The Del Monte Fish Company opened it in 1956, not far from the Point San Pablo Yacht Harbor. During its 15 years of operation, the 40-man crew boasted that they could reduce a humpback whale to oil, poultry meal and pet food in an hour and a half.

The whaling station's boats harpooned whales offshore along their California coastal migration routes, and towed them into San Pablo Bay, where they were pulled up a ramp by their tails with huge hooks. Station records indicate that in a typical year they brought in about 175 finbacks, humpbacks and sperm whales. Using very large knives, huge slabs of the whale were cut off and then cooked down in big pots, which produced the stench. That part of the Bay was usually filled with blood and brine and the remains of the processed whales.

It's hard to look back and imagine this today, but those were different times and the whaling station crew was working every day to make a living, just like everyone else. In the 1950s we didn't have the Environmental Protection Agency, Environmental Impact Reports, the Endangered Species Act or Marine Mammal Protection Act.

In 1971, however, Secretary of Commerce Maurice Stans signed the paperwork that ended whaling in the United States, and the Richmond whaling station was abandoned, ending 120 years of whaling along the California coast. After the site began leaking diesel oil into the bay, a hazardous waste remediation crew pulled out the underground tanks that fueled the station. A subsequent fire in 1998 destroyed much of what was left, leaving behind only charred pilings, memories and a few old photographs.

We have our own whaling history in Monterey Bay, although it began in the previous century. Captain John Davenport was the man who conceived of California shore whaling. He was a New Englander, originally from Rhode Island, where he had been half-owner of a 180-ton schooner that sailed between California and Hawaii from 1845 to 1852, whaling and trading. But whaling in those days of Captain Ahab and Ishmael was a difficult life. Whalers spent many months at sea, going after whales in longboats using hand thrown harpoons. Sometimes they didn't come back.

In 1852, John Davenport, newly married, moved west to California. Learning that humpback and gray whales were passing within view of Monterey, he soon recruited a dozen Portuguese men to hunt humpback and gray whales with hand lances and harpoons. He first whaled in the bay from a ship in 1853, but soon started whaling from small boats rowed out from shore. Thus began the enterprise known as shore whaling, also dangerous, but most times the men returned home every night.

Davenport's business proved to be economically marginal at best, although it soon attracted considerable competition. He sold out to a group of Portuguese, and returned to whaling from a larger ship - The Caroline E. Foote - in 1863, and spent two more years whaling along the California coast. His next venture was a shore whaling station at the mouth of Soquel Creek, where he operated unsuccessfully for two years. Interestingly, he blamed a large earthquake in the East Bay in 1865 for his lack of success at whaling.

Davenport then headed 15 miles up the coast to El Jarro Point, just north of the town that now bears his name. In 1866 he rented the land that is now called Davenport's Landing. He built a 450-foot wharf, which handled the shipping for several lumber companies and shingle mills. There is no evidence, however, that John Davenport ever built a whaling station in this location. This enterprise was also not that successful because the north coast swells kept tearing his wharf apart. Despite all of his ambition, John Davenport ultimately went bankrupt and moved into Santa Cruz.

While four sets of steel and concrete pilings from a pier can still be seen from the cliff across Highway One from the cement plant, these are not the remains of Davenport's wharf. This 2,600-foot long pier was built in the 1930s with the dream of shipping cement by sea from the north coast. This was a tough spot to tie up a ship, however, and after 15 years of marginal success, the pier was abandoned to the waves, and cement then left the county in trains and trucks. Transport of cement ceased completely when the cement plant closed in 2009.

Article 8

Whaling in the Bay

Whale gun loaded with a harpoon, c. 1900.
(courtesy: Scripps Institution of Oceanography)

Captain John Davenport starting shore whaling in California, but rowing out to harpoon whales from shore had its limits. It was hard and whalers could only work the waters out to about 10 miles from the beach. Even when they were successful in harpooning a whale, they then had to get a 15 to 30-ton animal back to the beach for processing, by rowing, or if they were lucky, with the aid of a sail. A fairly large percentage of the harpooned whales, especially humpbacks, were lost when they sank.

From the mid-1850s to about the 1880s, as many as 18 whaling stations were active along the length of the California coast, at Crescent City, Bolinas Bay, Half Moon Bay, Pigeon Point, Soquel, Monterey (2 stations), Carmel Bay/Pt. Lobos, Point Sur, San Simeon, Port Harford (Port San Luis), Cojo Viejo (Point Conception), Goleta,

Portuguese Bend and Dead Man's Island (San Pedro), and San Diego Bay (2 stations). By 1886 however, only five of these were still operating.

Following the discovery of oil in Titusville, Pennsylvania in 1859, which was refined to produce kerosene for lighting and heating, the value of whale oil and the economics of whaling declined quickly. While the shore whalers certainly reduced the number of whales along California's coast, it is not clear whether or not the reduction in their numbers was a significant factor in the decline of the industry.

In the early 1900's, some new technologies like steam powered chase boats and the harpoon cannon, led to a brief resurgence in whaling, including a short-lived Norwegian-owned whaling company that operated at Moss Landing between 1919-1926.

Another retired ship captain, Charles Moss, brought his family from Texas in 1866 to build a homestead in California. Partnering with Portuguese whaler Cato Vierra, Moss built a 200-foot long wharf on the sand spit directly west of the present Sandholdt Bridge to create a commercial pier. The wharf was soon busy, attracting whalers, fishermen, and salt pond operators to Moss Landing Harbor. The port's traffic was further boosted by gold rush fever when Moss Landing Harbor began exporting goods like potatoes, lumber, and sugar beets to a booming San Francisco. Extraction of salt from evaporation ponds and oyster farming soon followed.

Though Charles Moss eventually sold his holdings in the area to the Pacific Coast Steamship Company, residents seem to have respected him enough to adopt his name for the area. The 1906 San Francisco earthquake took its toll on Moss Landing, seriously damaging the old wharf and much of the town's infrastructure.

In 1919, the California Sea Products Company and a Norwegian whaler, Captain Frederick Dedrick, selected Moss Landing as the site for a commercial whaling factory. Whaling at Moss Landing began in earnest again. Two new steam driven boats with bow mounted harpoon cannons would make whaling in the bay a whole lot easier for the whalers, but a whole lot more dangerous for the whales.

Whale being flensed at Moss Landing Whaling Station, c. 1900.
(courtesy: Scripps Institution of Oceanography)

The deck-mounted harpoon cannon, and also harpoon guns held by whalers at the shoulder, were both used to get a line on the whale. This was usually followed by a bomb lance shot from a gun, or in some cases, the harpoon and bomb gun were mounted together on a pole and thrown as one would an ordinary harpoon. When the harpoon entered the whale a certain distance, it engaged a trigger that fired the explosive. While whales stood a reasonable chance of evading the earlier shore whalers, the introduction of steam powered ships and the harpoon cannon and bomb gun led to much higher take rates. An interview with one of the harpoon gunners who worked out of the Richmond station, the last whaling facility on the west coast, described the injuries that befell many of the gunners as well. Whaling seems to have always been a risky business, no matter what tool or technology was used.

Data from logbooks of whaling stations at Moss Landing (1919-1922 and 1924) and Trinidad (1920 and 1922-1926) record the taking of 2,111 whales, including 1,871 humpbacks, 177 fin whales, 26 sei whales, 3 blue whales, 12 sperm whales, 7 gray whales, and 15 others. This works out to be about 80 whales every year for each of these two stations.

Sandy Lydon's history site explains how the arrival of each whale at the new Moss Landing whaling station was initially a spectacle that attracted large numbers of people, at least large numbers for Moss Landing at the time. After a year or so, the novelty wore off and the stench from the processing plant drove people away. When winds blew from the south, the odor of a whale being boiled down could be smelled all the way to Santa Cruz.

The Moss Landing plant used just about every part of the whales they brought in. The blubber was turned into oil used by soap manufacturers; the meat was converted to chicken feed, and bones were ground into bone meal.

Within several years of the whaling resurgence in Monterey Bay, the new tools had proven to be very effective and the whales were become both scarce and wary. By 1924 the last whale had been brought ashore and the Moss Landing site was closed, bringing an end to whaling in Monterey Bay.

The Richmond whaling station opened for business in 1956, thirty-two years after whaling in Monterey Bay ended. Perhaps the whale population had recovered sufficiently for whaling to become profitable again. However a short fifteen years later, federal protection closed down this last west coast whaling factory, ending 118 years of California whaling. Today, whale populations have slowly rebounded, but are still not believed to be anywhere near the numbers that passed along our shores prior to human intervention.

Article 9

West Cliff - A Local Treasure

Vue de L'eau streetcar and station on
Garfield Avenue (now Woodrow), Santa Cruz, c. 1920.
(courtesy: Harold J. Van Gorder Collection)

Whether walking, jogging, biking, pushing a baby stroller, or a century ago, bouncing along in a wagon, West Cliff Drive has always been one of those places that make Santa Cruz special. Much of the oceanfront along East and Opal Cliffs are covered with homes, leaving only a few areas where you can walk, bike or drive along the edge of the Pacific. We are fortunate that the earliest wagon road on Santa Cruz' west side followed the bluff edge for virtually its entire three-mile length from Cowell to Natural Bridges. Often called the Cliff Road, this was one of the city's earliest, and at times, dustiest visitor attractions.

It's probably a good thing that only a single house was ever built on the ocean side of West Cliff, simply because waves have gradually moved the cliff edge back over the years. Those who walk along the path regularly will notice changes from year to year. New sea stacks form where bridges and arches have collapsed, with Natural Bridges being perhaps the most photographed example; but there have been many others.

The winter storms, particularly in big El Niño years like 1982-83 and 1997-98, wreak the most havoc. Erosion and damage are the greatest when large waves arrive at times of high tides and elevated sea levels from warmer El Niño waters.

During January, February and March of the 1983 winter, the coast of California sustained about $230 million in losses to private and public property (in 2013 dollars). Waves overtopped the bluff along West Cliff and took their toll on Lighthouse Point, the bike path, and also closed some of the parking areas. Collapse and failure of the cliffs is an episodic process, however, and you might have to stand out there a long time to actually see a large chunk of cliff fail or an arch collapse.

The end of Woodrow Avenue (formerly Garfield) has seen some dramatic changes over the past century or so. Woodrow has a distinct median strip, which accommodated one of the city's original streetcar routes. The tracks ended on West Cliff Drive at a picturesque Victorian gazebo known as the Vue de L'eau, or view of the water. The Santa Cruz, Garfield Park and Capitola Electric Railway built it in 1891 as an observatory, waiting station and depot.

In 1905, the Norris and Rowe Circus bought the Vue de L'eau and the surrounding six acres for use as its winter quarters. For a short time in the early part of the last century there was also a marine museum on the inland side of West Cliff Drive at Woodrow.

Sadly, the gazebo burned to the ground in 1925. Although the streetcar tracks are long gone, the city has recently put some of the original streetcar wheels on display at the corner of Woodrow Avenue and West Cliff Drive, along with some historic photographs of one of the old streetcars rumbling down Garfield towards the Vue de L'eau.

A short distance upcoast of the gazebo was an often-photographed arch, variously known as Arch Rock and also Crown Arch (for a while it had a flat cap or crown of weaker sandy terrace deposits). Lots of old photographs and postcards of Crown Arch remain, and as the years progress you can see the crown disappear, the arch gradually thin, and finally sometime in the 1970's, it collapsed. This also looks to be the site of a unique double arch, which was also a common site for photographs at the turn of the last century.

As we put the old photographs together for the "Santa Cruz Coast-Then and Now" book a few years ago, finding the exact location of Crown Arch gave us as much trouble as any other West Cliff Drive images. There are some fingers of rock that extend out into the water near the end of Woodrow, which seem to be the most obvious locations for the base of the old arch. But looking at the photographs with Woodrow Avenue and the gazebo, and then comparing these with photographs with the gazebo and Crown Arch, the location that seemed the closest were always difficult to match up with the old photos.

So we went back a dozen or more times, always thinking that on the next visit it would all become clear. We did our best in the book but we were never completely sure we were standing in the right place. And in fact, the right place may have been eroded away in the intervening years.

Stop here on your next walk with the old photograph in hand (you could use the one in this article, or if worse comes to worse, you could borrow the book from someone) and try to find it yourself. Let me know when you feel you have figured it out.

Article 10

View of the Santa Cruz Waterfront - 1906

Panoramic view of the Santa Cruz waterfront, 1906.
(photo: George Lawrence, courtesy: Library of Congress)

A Midwestern inventor and photographer, George Lawrence, took the first aerial view of the Santa Cruz coast in 1906 from a camera carried aloft by a system of kites. This rarely seen panoramic photograph covers the entire city and shoreline from Cowell Beach to nearly Pleasure Point and provides a remarkable and unique perspective from over a century ago.

Lawrence was born in Illinois right after the Civil War. He first worked in a buggy shop where he invented a way to attach iron rims to wooden wheels. In 1891 he opened a photography studio in Chicago where he began to channel his creative energies. He developed what was known as flashlight photography, which was used until flashbulbs were invented many years later.

He quickly became an innovator in the expanding field of photography and developed the slogan: "The Hitherto Impossible in Photography is Our Specialty". In 1900 he built a massive camera weighing 1400 pounds in order to take a photograph of a large locomotive for the Paris Exposition of 1900, which won "The Grand Prize of the World". Not many people can make that claim.

Soon after the turn of the century, George Lawrence became intrigued with the idea of taking photographs from an aerial perspective. Initially he used ladders and high towers but these were pretty limiting and a bit dangerous. Airplanes were still in their infancy, however, and the Wright brothers' first successful flight at Kitty Hawk, North Carolina, didn't happen until 1903.

In order to get the elevated views he wanted, George devised another approach, using a hot air balloon. In 1901 he began shooting aerial photographs from a flimsy cage attached to a tethered balloon. Over the next several years he took a large number of photographs of cities across the country, conventions, sporting events and even western wilderness areas.

But in 1901, while 200 feet above Chicago, the cage with Lawrence in it broke loose from the balloon. Fortunately, telephone wires broke his fall and he landed unharmed. Some months later he fell a second time. At this point he decided that maybe he ought to figure out a safer way to take photographs from the air.

So George began to develop a system of kites to carry bigger and bigger cameras. He used a large plumb bob or weight to keep the kites from spinning around, and used an electric current through the kite's main tether to trip the shutter. In order to get the greatest panoramic view, he developed a camera with a lens that could swing in an arc on a semi-circular track. This is all pretty amazing if you stop and think about it. I was never that great at flying kites, but Lawrence used as many as 17 at a time to lift his massive camera, which would move on a track hundreds of feet in the air.

Perhaps his best-known photograph is a panoramic view of San Francisco taken in May 1906, just a few weeks after the devastating

San Francisco earthquake and fire. With considerable effort he was able to coax his array of kites up to an altitude of 2000 feet to capture virtually the entire city, including the Golden Gate, on a single 17-by-48 inch piece of film.

In that same year, George Lawrence traveled to Santa Cruz and launched his kites from the West Cliff Drive area, not far from the lighthouse, and captured the details of the city and the coastline.

High-resolution digital copies of Lawrence's 1906 aerial photographs of San Francisco and Santa Cruz can be downloaded from the following Library of Congress web-site: http://memory.loc.gov/cgi-bin/query/d?pan:20:./temp/~ammem_uS6V: The Santa Cruz image is No. 53, and San Francisco is No. 110. You can click on the image to download high-resolution copies. And if you happen to have a large format printer in your home office, printing them at their original scale, 48 inches across, produces a great wall covering or conversation piece.

There were three wharves at that time, all gone or replaced now. The one in the foreground is the Cowell Wharf, which was at a higher elevation and left the cliffs at the end of Bay Street, right between where the Dream Inn and the Sea and Sand stand today. Nearys Lagoon shows up as a large wet swampy area.

The second largest house built in Santa Cruz up until that time, the Sedgewick Lynch house, is visible on the terrace between the two longest wharves, just inland from West Cliff. It was constructed at a cost of $12,000 at the time and has recently been renovated as The West Cliff Inn, right across the street from the Dream Inn. There are a few other recognizable structures but it's a very different waterfront today, and George Lawrence managed to capture it all from a string of kites over a century ago.

Article 11

The Wharves of Santa Cruz

Close-up view of the Santa Cruz waterfront, 1906.
(photo: George Lawrence, courtesy: Library of Congress)

The Santa Cruz waterfront has seen a number of piers or wharves come and go over the past 165 years. Before the construction of the first structure in 1849, passengers and produce had to be ferried out through the surf in small rowboats and lumber was actually floated out to ships waiting offshore. As you can imagine, this wasn't a terribly efficient process and also couldn't be done when there was high surf.

The first wharf was really just a ramp. It was built in 1849 at the end of Bay Street by Elihu Anthony, and was initially used as a chute to load potatoes onto ships. Bay Street was the primary access to the chute so that all of the lumber, lime and other cargo didn't have to be transported through downtown. A chute from the end of Bay Street also had the advantage of being at a higher elevation than the beach front so things could be moved downhill a whole lot easier. Anthony also built the first bridge in the county to provide access to his wharf off West Cliff Drive. The last Howe strut bridge in Northern California still straddles the outlet from Nearys Lagoon just behind the new Monterey Bay National Marine Sanctuary Exploration Center.

A few years later in 1853, Jordan and Davis Limeworks took over the chute. Henry Cowell bought out Jordan and Davis in 1867 and the wharf became known as Cowell Wharf. The photograph that George Lawrence took from the kites in 1906, and enlarged in this article, shows Cowell Wharf and also the Henry Cowell Limeworks warehouse at the end of Bay Street, where lime was stored prior to shipping. Cowell Beach probably originally received its name from the wharf.

The 1906 photograph also shows a second wharf, a little to the west of the location of the present Municipal Wharf, which was built by David Gharkey in 1856. The South Coast Railroad bought this wharf twenty years later and it soon became known as the Railroad Wharf. Tracks were extended to the end and freight and passengers could then be transferred directly to or from ships.

During the Civil War, another wharf was built a short distance to the east. This wharf was built to serve the California Powder Works and started on Beach Hill at the intersection of Second and Main streets. The Powder Works, located on the San Lorenzo River two miles upstream from the city, was the only blasting and gunpowder manufacturing plant west of the Mississippi River. They shipped in nitrate from South America and then shipped blasting and gunpowder out. What isn't clear is whether the powder was shipped to the Union or the Confederate armies, both, or neither, and what Santa Cruz added to the process other than a place without too many people around in case there were any accidental explosions, which there were.

This structure became known as the Powder Company Wharf. Between 1877 and 1882 it was connected to the Railroad Wharf by a cross wharf. In 1883, however, both the Powder Company Wharf and the cross wharf were demolished as most of the gunpowder was now shipped out by rail.

In the George Lawrence photograph, a 3rd and shorter pier is shown directly in front of the Boardwalk. This was built in 1904, in part to carry a pipe for seawater to fill the new plunge pool with fresh salt water every day. It was initially known locally as the Electric Pier because of its night lighting, and then later as the Pleasure Pier. Small boats took people from the pier to an amusement ship, the Balboa, anchored just offshore. When the saltwater plunge was replaced by a miniature golf course in the mid-1960s, this pier was demolished.

In 1907, a year after Lawrence took this early aerial photograph, large winter waves destroyed Cowell Wharf. All of these early wharfs were relatively short, however, which limited the water depths they reached, and therefore the size of the vessels that could be accommodated. In order to serve larger and deeper draft vessels, the first Municipal Wharf, 3000 feet long, was constructed in 1914. This is essentially the same structure that exists today, although lots of pilings and decking have been replaced over the years.

The outer end of this new Municipal Wharf was turned more to the south in order to meet the North Pacific swells refracting around Lighthouse Point more head on. Some 2000 70-foot long Douglas fir pilings were driven about 20 feet into the seafloor to support the wharf, the railway line, which was moved from the Railroad Wharf, and a freight warehouse. The old Railroad Wharf was used to support a sardine cannery before being torn down in 1922.

There is a rich local history of the Italian fishing community that started on the Railroad Wharf in the 1870s and then relocated to the Municipal Wharf in the 1920s. Their unique fishing boats, the davits for lifting them out of the bay, and the fishermen working on their nets and cleaning their fish added much to the city's cultural environment.

The fishermen would occasionally display unusual animals brought up in their nets, and there was even an early aquarium on the wharf for visitors to enjoy. When the small craft harbor was constructed in the mid-1960s, this all changed as the fishermen and their boats were relocated. What remains on the Municipal Wharf today are the fish markets and restaurants, bearing the names of the original Italian fishing families, Stagnaro, Castagnola and Carniglia.

Much of the information on the history of the wharfs in Santa Cruz is drawn from a detailed summary by Frank Perry, Barry Brown, Rick Hyman and Stanley Stevens. Peter Nurkse originally brought George Lawrence's photographs to my attention and has done an interesting study and explanation of the original 1906 photograph: http://www.santacruzpl.org/history/articles/182/

Article 12

Finding Monterey Bay

In June 2012, Sandy Lydon and I led another group of 40 dedicated hikers on our 4th annual Monterey Bay walk from New Brighton beach to the breakwater in Monterey. The bay is a big curious bite out of California's coast, easily recognized by all of us, but one that wasn't so obvious to early explorers.

While there are a number of perfectly curved bays along the state's Pacific edge, uncoiling from north to south like giant abalones - Half Moon Bay, Bodega Bay, Drakes Bay, and Stinson Beach, to name a few - Monterey Bay is unique in having smoothly curved beaches at both ends.

The bay appears to have been first carefully explored and mapped by Sebastián Vizcaíno in 1602, over 400 years ago. The Spanish Viceroy in Mexico City, the Count of Monte Rey, appointed Vizcaíno as the general - in - charge of an expedition to locate safe harbors in Alta California for Spanish galleons to use on their return voyage from the Philippines to Acapulco. He was given the authority to map in detail the California coastline that Juan Rodriguez Cabrillo had first explored 60 years earlier.

As he progressed up the coast he named and also renamed many of the prominent features, San Diego, Santa Catalina Island, Santa Barbara, Point Conception, Carmel Valley and Monterey Bay that had been earlier named by Cabrillo. Vizcaíno sailed as far north as Cape Mendocino, making detailed charts of the coastline.

It was a difficult voyage with most of Vizcaíno's crew suffering from scurvy because of lack of vitamin C. Sixteen sailors had died by the time he reached what is now Monterey Bay. While he gave new names to many places, the only place he apparently actually explored in detail was Monterey Bay. In fact, he mapped the bay's coastline so carefully that his maps were used for the next 200 years.

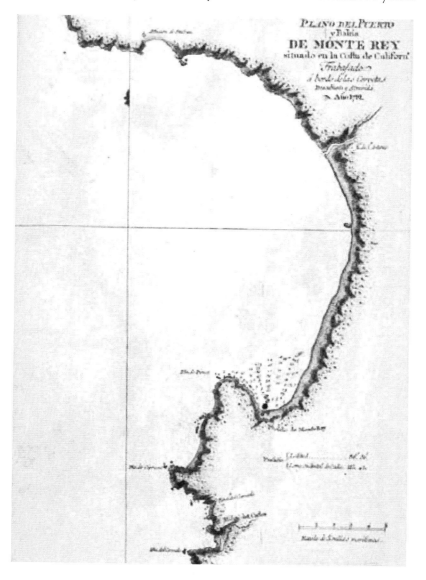

First map of Monterey Bay by Sebastian Vizcaíno
(from the book: Relacion de Viage Hecho por Los Goletas Sutil y Mexicana,
by Villa Alcalá Jorge, 1802)

Vizcaíno reported in his logs that Monterey Bay was a safe harbor, "sheltered from all winds". As was discovered in subsequent years, however, Monterey Bay really has no natural harbors and considerable effort was expended in the last century to build harbors sheltered from the wind and waves at Santa Cruz, Moss Landing, and Monterey.

Although Vizcaíno spoke highly of the California coast and Monterey Bay as a good port for Manila galleons, he was not allowed a return visit and Alta California was ignored for over a century and a half. European issues as well as a perceived threat to the occupation of the California coast from England and Russia also diverted Spain's attention at the time.

In the spring of 1769, two expeditions, one commanded by Captain Gaspar de Portolá, and a second by Captain Fernando Rivera, set out from Baja California to prepare for the military occupancy of Alta California. One very important objective was to reach and explore Vizcaíno's famous Monterey Bay harbor. The threat of Russia and England along the coast was much stronger at this time, which provided an additional catalyst for the expedition.

While today you can jump in your SUV in San Diego, get on I-5 and 101 and drive to Monterey in a day. The Portolá expedition took two and a half not always comfortable months. They left San Diego on July 14, 1769, and headed north along the coast. After passing Santa Barbara on August 19, they reached the southern end of the rugged and essentially impassable Big Sur coast on September 13, and were forced to make a difficult detour inland through the Santa Lucia Range to the Salinas Valley. Following the river downstream towards Monterey Bay they reached an area between Marina and Salinas on October 1 where they camped.

Portolá and a small group of soldiers headed towards the river mouth. They climbed a low sand hill, known today as Mulligan Hill, near the mouth of the present day Salinas River, but failed to see or recognize Vizcaíno's harbor "sheltered from all winds". Although October is typically some of our best weather, perhaps Portolá's view was obscured by fog. Mulligan Hill is actually a large, now vegetated sand dune, visible from Highway 1, and which today provides a good view of the entire bay. The dunes along this stretch of central and southern Monterey Bay shoreline bears evidence of a long history of strong onshore winds.

A scouting party explored the Monterey Peninsula, but didn't find the protected harbor there either. They were also eagerly anticipating meeting a ship, the San José, which was carrying supplies, but which unfortunately never arrived. A number of men on the expedition were now incapacitated from scurvy and were being carried on litters, which didn't help matters or morale.

They continued north, reaching the Pajaro River, which they named after a large straw-stuffed bird that was left behind in a deserted Indian settlement. After stopping at Pinto Lake, the expedition reached the area of present day Santa Cruz on October 18, 1769. Finally, after a rugged and very difficult journey up the North Coast, they sighted San Francisco Bay from San Bruno Mountain on October 31.

While San Francisco Bay was soon recognized as one of the greatest natural harbors in the world, Portolá apparently thought little of San Francisco Bay and was disappointed in not being able to find Monterey Bay's harbor "sheltered from all winds".

The expedition turned around, and headed back towards the Monterey Peninsula. Reaching the Carmel Bay area, the expedition's pack animals found lots to forage on but the explorers themselves were reduced to eating seagulls and pelicans. Local Indians may have saved them by bringing them ground corn and seeds.

They continued to believe that the San José, the resupply ship they were expecting, would find them. In one last attempt to communicate, they erected two wooden crosses on low hills above the beach, with notes buried at their bases, hoping that the crew would see them and come ashore.

It was now December, however, and the weather was deteriorating so they made the decision to head south. The remaining party, minus several deserters, reached San Diego on January 24, 1770, six months after their departure. The ship that the expedition had been looking for had been forced to return to port in Mexico for repairs shortly after departing. It never reached Monterey Bay, and in fact, was never heard from again.

    Portolá was a very determined soldier, however, and he returned that spring and this time he recognized Vizcaíno's Monterey Bay, although, as it is today, it is never really "sheltered from all winds". The Santa Cruz corner of the bay is sheltered from northwest winds and the Monterey corner is protected from southwest winds, but there is no natural harbor that is protected year round.

Article 13

Coastal Place Names

A number of faithful readers responded when I asked about suggestions for future articles. After five years of columns I thought it would be interesting to hear what you wanted to know more about. While some readers requested topics I had covered in earlier columns, others had new ideas. I am trying to get to each of those, but as you probably realize, I often get side tracked by some current ocean or coastal issue.

One idea I found interesting was the source of many of our local coastal place names, simply because whether we are new to the area or old timers, there are lots of interesting names that few of us know origins for. Names or places like Panther Beach, Black Point, and Hidden Beach, to name a few.

This column, and a few more to come, are in response to this suggestion. I am deeply indebted to a wonderful source of information on all things Santa Cruz: Santa Cruz County Place Names by the late Donald T. Clark. This author is the authority and this book is one to be trusted.

Don Clark was the founding UCSC librarian, and reportedly the first of founding Chancellor Dean McHenry's appointments, back a half a century ago now, in 1962. He was a historian in addition to being a librarian. In the Forward to the book, Sandy Lydon, a close friend and local historian, writes that Don Clark dismissed himself as a "trivia buff", but spent 25 years researching the origins and history of every conceivable place name in the region (2,300 of them to be exact), and this is the Rosetta Stone for the county.

The book is out of print but copies can be found on-line, and while it is not exactly a novel, it makes for fascinating and illuminating reading. And when there is a disagreement among earlier references or historians (historians rarely agree on anything), Don Clark provides us with all the information and lets us make up our own mind.

Being sort of an organized person, I thought it made sense to proceed along the coast in one direction or another, rather than use a random approach. So I made the decision to traverse the coast from north to south, from Waddell Bluffs to the Pajaro River, the northern and southern boundaries of the county coastline.

Lets head north to Waddell Bluffs, Beach and Creek. Who was Waddell anyway? Well, William Waddell was originally from Kentucky and worked his way across the continent as did many others in the middle of the 1800s, arriving in Santa Cruz County in 1851. At one time or another he built and operated four different lumber mills, with the last mill along the creek now named after him.

In 1861 he completed a sawmill, a five-mile long tramway using horse drawn cars to get the lumber to the coast, and a 700-foot long wharf, which extended out from the bluff a short ways north of where the Año Nuevo State Park Visitor Center is today. A small settlement, Waddell's Landing developed and the wharf lasted until storms destroyed it about 1880.

Unfortunately, while hunting deer in 1875 he was attacked and severely mauled by a large Grizzly bear. William Waddell died of his injuries several days later at age 57.

Article 14

Exploring the North Coast Names & Places

Scott Creek, Santa Cruz County.
(© 2012 Kenneth & Gabrielle Adelman,
California Coastal Records Project, www.californiacoastline.org)

I introduced Donald Clark's wonderful book, Santa Cruz County Place Names, in my last column. Mike Clark, Don's son, let me know that the book was revised and reprinted, and is available.

Getting back to the North Coast, but heading south, Greyhound Rock appears as large mass of Santa Cruz Mudstone, which in the summer months is usually connected to the shoreline by a sandspit. A spit connecting a small island to the coast is known as a tombolo. There are many times, however, either at very high tides or during the winter months, when the spit is submerged, stranding unsuspecting adventurers on the rock.

Why Greyhound Rock? Donald Clark's book states that a Santa Cruz pioneer, Thomas Majors, explained how it got its name - because it "looks like a hound, you know". Most people have great difficulty recognizing a hound and simply see a large lump of rock. Some even see a whale.

A mile south of Greyhound Rock is a place with a strange name but which is little known today, China Ladder. A few Chinese fishermen lived in small shacks at the top of the high steep cliffs here in the late 1800s. They used a rope that led to a ladder, which they descended to the shoreline, where they obtained abalone that were then dried and sold to the Chinese market.

Scott or Scotts Creek, depending upon whom you talk to, was named after Hiram Scott, originally from Maine, who arrived in Monterey by ship in 1846. Being from a large family he was forced at an early age to support himself. One of the few opportunities for a young man at the time was that of a sailor, so he went to sea. Through hard work he was successful, and rose to become a 2nd mate, but soon realized this wasn't the life for him.

Jumping ship in Monterey he made his way to Santa Cruz, where, somewhat surprising for a man who disliked life at sea, he began building a boat along what is now Main Beach. The California Gold Rush took him north and east in 1848, where he did well in the mines. He ultimately settled in Stockton where he built a grand hotel and also started a ferry service.

Santa Cruz drew Scott back, however. He bought the San Augustin Rancho in the valley that now bears his name for $25,000, and settled down to raise potatoes, hay and fine horses. His continued financial success led him in 1852 to buy a second ranch up the coast, a portion of Rancho Agua Puerca (dirty or muddy water) y Las Trancas (bar or barrier), near Scott Creek.

The windswept beach at the mouth of Scott(s) Creek is known as Scott(s) Creek Beach. The strong onshore winds from the northwest blow sand into dunes that used to migrate inland across the route of present day Highway 1. The dunes just above the beach were formerly the site of a tank farm. In the 1940s and 1950s oil was pumped through a pipeline from offshore into a series of large storage tanks for the cement plant. In later years the dunes were the site of a mushroom farm.

The wide marine terrace between Scott(s) Creek Beach and Davenport Landing is known as El Jarro Point. The name appears to have come from the original 1839 Mexican land grant, Rancho El Jarro. The source of the name, El Jarro, which means jug or jar in Spanish, seems odd, and has to my knowledge, never been understood or explained.

This oceanfront terrace hit the front pages of the Sentinel in 1969 when PG&E announced that they had taken out a lease on 6,800 acres of Coast Dairies and Land property with the intention of building what would have been the world's largest nuclear power plant at El Jarro Point. This proposal, as much as any other single event, ushered in the environmental movement in Santa Cruz.

Article 15

Captain Davenport's Landing

Captain John Davenport, c. 1852.
(courtesy: Alverda Orlando)

As a kid heading north from southern California in the 1950s on summer camping trips in an old station wagon, I recall passing through a small town on the coast somewhere south of San Francisco that seemed completely enveloped in gray dust. After arriving in Santa Cruz at the newly opened UCSC campus in 1968 and while exploring the coast to the north, I rediscovered Davenport, still covered in gray cement dust.

There's lots of history in Davenport, including how it got its name. Davenport's Landing, at the mouth of Aqua Puerca Creek (muddy or dirty water) about a mile and a half north of the present town, was named after Captain John Davenport, a sailor who arrived in California from Rhode Island in 1849.

Originally settling in Monterey, Captain Davenport is recognized for being one of the first, if not the first, to begin shore-based whaling. From Monterey he moved north to Moss Landing, where he helped establish a major whaling station. In 1867 he moved his family to Soquel Landing (Capitola), helping F.A. Hihn build the Capitola wharf. He then headed farther north to El Jarro Point, where he constructed a 450-foot long wharf.

Lumber, tanbark and cordwood seem to have been the main products shipped from Davenport's Landing. A small community developed around the small cove and at its peak in the 1870s and 1880s had three hotels (the Bannister, the Davenport Landing Exchange and La Stella del Mar), two general stores, a blacksmith and butcher shop, a shipyard and several houses. I'm not sure what happened to all those buildings, but it doesn't appear today that a trace remains of any of them.

The town of Davenport actually came along several decades later and acquired its name in 1905. The large deposits of marble and shale in the adjacent mountains led William Dingee, an eastern cement entrepreneur, to develop the Santa Cruz Lime Company along the Coast Road at San Vicente Creek. I think the community is fortunate that it was named after a whaler rather than a cement baron.

For a little over a century, cement was produced at Davenport and shipped out by rail, truck and for a while, even by sea from a pier directly opposite the cement plant. The CEMEX plant was closed a few years ago, however, for economic reasons.

One mile south of Davenport, Liddell Creek has eroded a canyon through the marine terrace to the ocean and formed a large pocket beach known as Bonny Doon Beach. Higher in the hills, Liddell Creek has provided an important source of very high quality drinking water for the city of Santa Cruz for over a century.

George Liddell was an English contractor and civil engineer who arrived in California in 1850, like so many others. Within a year he had settled in the Santa Cruz Mountains and built both a water mill and a steam-powered sawmill on the creek, which is named after him.

Today Bonny Doon Road leaves Highway 1 at the mouth of Liddell Creek and winds its way up the mountain to an area known today as Bonny Doon, but historically as Battle Mountain, because "...the rocks are piled in confused splendor and the imagination easily conjures up ancient castles..." (written in 1875).

Liddell Creek, like virtually all of the other North Coast streams between Santa Cruz and Davenport, no longer flows freely to the shoreline, but passes through a hand-dug tunnel through the cliffs to the beach after crossing under both the railway and Highway 1.

One of the most spectacular and geologically fascinating beaches along the entire north coast lies a little less than a mile farther south, known historically as Yellow Bank Beach. Since the 1960s, however, some people with enhanced visual acuity have recognized the outline of a large cat-like creature in the cliff face as you descend to the beach. Those folks often refer to the area as Panther Beach. Next column, why Yellow Bank Beach?

Article 16

Yellow Bank Beach

Yellow Bank Beach, Santa Cruz County.
(© 2006 Gary Griggs)

My last column probably left thousands of readers hanging at Panther Beach, eagerly awaiting the next episode of adventure along the North Coast. Although known today by most visitors as Panther Beach, historically it was known as Yellow Bank Beach, which begs the question, why Yellow Bank?

The Yellow Bank Dairy along the North Coast is mentioned in the old records of Santa Cruz County from 1887 as one of a number of coastal dairy farms. Theses dairies started at the Natural Bridges Dairy on the south and extended as far north as present day Davenport.

Seven miles north of the Santa Cruz City limits, you cross what today is called Yellow Bank Creek. On the old maps it's called Respini Creek, after Jeremiah Respini, a farmer from Switzerland. He ran a dairy in the small valley near the mouth of the creek, named - you guessed it - the Yellow Bank Dairy.

From Highway 1 you can look down as you cross the highway embankment into that little narrow valley and see an outcrop of yellow or golden colored rock on the south side. The exposure is actually part of the cut that was made for the Old Coast Road before the valleys were all bridged with fill for today's Highway 1. And while you can't see the beach on the opposite side of the highway because of the railroad embankment, the vertical cliffs there consist of this same gold colored rock; only the exposures are much larger.

The bedrock exposed in the seacliffs from West Cliff to Waddell Bluffs, and in the road cuts along Highway 1, is known geologically as the Santa Cruz Mudstone. It's a tan colored, and not particularly exciting geologic formation, devoid of any visible fossils. The mudstone was deposited in an ancient sea that covered this area some 5 to 10 million years ago.

The yellow or gold colored rock is actually sandstone, and it was intruded or injected into the mudstone like soft toothpaste from an underlying formation millions of years ago. A mining geologist, John Newsom, who had come west from Indiana in 1901 to study geology at Stanford University, first recognized the unique nature of these sedimentary intrusions along the coast while mapping the area between 1901 and 1902.

These features that gave Yellow Bank Creek, Dairy and Beach their names, at least until the panther was first seen on the cliff face, are believed to be the largest exposed sedimentary intrusions anywhere in the world. Not many places can claim that.

While some intrusions are a gold color due to oxidized iron, there are others exposed in the seacliffs that are a blue-grey color.

Looking up Majors Creek canyon, two miles to the south, you can see massive, near vertical cliffs that form the edges of the canyon and which consist of massive, black, asphaltic sandstone intrusions. Similar but smaller intrusions cut through the mudstone of the seacliffs and road cuts from Wilder Ranch to Greyhound Rock.

These sand bodies are called sills when they are more or less horizontal and dikes where they are near vertical. Igneous dikes and sills are very common in the geologic record and were formed when hot magma under pressure was intruded into the overlying rocks. It was this process that concentrated gold in the veins of the Sierra Nevada.

Sedimentary intrusions, however, are much less common and require a special set of conditions. We need a deposit of loose clean sand that is saturated with a fluid, usually water or oil. Then, due to either a large earthquake, or the pressure of hundreds or thousands of feet of overlying material, the sand can liquefy, essentially converting it to a fluid, like quicksand.

The overlying pressure can squeeze this now liquefied sand upward through cracks or weak zones into the overlying rocks where it gradually solidifies. Think about jumping on a huge tube of toothpaste the size of an Olympic swimming pool. And that is how we created the sandstone intrusions that gave Yellow Bank Beach its name.

ARTICLE 17

Sand Hill Bluff

Sand Hill Bluff, Santa Cruz County.
(© 2009 Kenneth & Gabrielle Adelman,
California Coastal Records Project, www.californiacoastline.org)

Sand Hill Bluff, five miles north of the city limits, is a unique geographic feature, labeled on the USGS topographic map of the North Coast, but probably missed by most travelers passing by on Highway 1. It's distinct in being one of the only areas along the coast north of Santa Cruz where there is any relief on top of the normally flat marine terrace.

Sand Hill Bluff was first called out in the 1889 Pacific Coast Pilot as a location along the shoreline where a large accumulation of white sand could be seen from ships offshore rising above the tan colored mudstone cliffs that make up the coastline.

It has retained this name to the present, although much of the surface of the white sand is now covered with vegetation. The area is also a kitchen midden, as evidenced by the fragments of abalone and other marine invertebrate shells left behind by Native Americans who occupied the area several thousand years ago and harvested shellfish from the intertidal zone. The area is now fenced off and protected although is surrounded by a boardwalk for hikers.

Why is there a sand pile sitting on top of the terrace at this particular location? Interestingly, the pile of sand is actually an old dune, but one that is no longer connected to its original source of beach sand.

In order to build a large sand dune, we need an ample supply of beach sand, a wide beach so the sand can dry out between tidal cycles, a dominant onshore wind, and then an area of low relief directly inland from the beach where the sand can accumulate.

Today, the old dune is separated from the shoreline below by a 60-foot high vertical cliff. This lack of connection indicates a earlier condition when sea level was tens of feet lower than today and the shoreline was thousands of feet farther offshore. The seacliff was very low or non-existent at that time, perhaps like the north shore of Año Nuevo Point today, where sand could be blown inland to form dunes. The dunes were cut off as sea level rose and the shoreline migrated landward, with waves gradually eroding a sea cliff into the flank of the Santa Cruz Mountains.

Immediately inland from Sand Hill Bluff is the old community of Majors, a small settlement of homes along a short section of the Old Coast Road on the ocean side of Highway 1 between Majors and Laguna creeks. Majors was a former stop on the Davenport to Santa Cruz section of the Southern Pacific Railroad, formerly knows as Enright.

So why Enright? Well, we have to go back to the middle of the 1800s and the arrival of James Enright, a native of Ireland, who crossed the Great Plains in 1846 and arrived in Santa Cruz not long afterwards. Enright was known as one of the most prominent dairymen of early Santa Cruz County and built a ranch of a thousand acres along this stretch of coastline. He and his ranch were known for their neatness and thrift.

Enright went on to serve on the County Board of Supervisors and was also one of the organizers of the Santa Cruz Oil Company, which attempted to extract oil from the asphaltic sandstone that underlies the uplifted terraces between Majors and Laguna creeks.

While some modest amount of oil was pumped out through heating the rocks at depth, it wasn't a profitable venture. James and his son Joseph Enright, also developed what was likely the first asphalt mine in the area on the family ranch, known as the Enright Mine. Between 1888 and 1914, an estimated 614,000 tons of asphaltic paving material, worth about $2.3 million, was extracted from the north coast. Production was intermittent after the 1920s, with the last of the quarries ceasing operations in the 1940s.

Article 18

Early North Coast Ranchers

Red, White and Blue Beach, Santa Cruz County.
(© 2013 Kenneth & Gabrielle Adelman,
California Coastal Records Project, www.californiacoastline.org)

Immediately inland from Sand Hill Bluff is a short section of the Old Coast Road, now bypassed by Highway 1. There is a small settlement of perhaps a dozen homes here that goes by the name of Majors, which was a former stop on the Southern Pacific Railroad.

The next major drainage to the south is known as Majors Creek, which has been one of the city's important North Coast sources of high quality water for over a century. This steep rugged canyon, visible just inland from Highway 1, empties into what was originally named Majors Beach, but for nearly 40 years was better known as Red, White and Blue Beach.

Joseph Ladd Majors, originally from Tennessee, came across the Santa Fe Trail and arrived in California in 1834, settling in Santa Cruz the next year. Majors married Maria de Los Angeles Castro and became naturalized as a Mexican citizen. He was soon granted Rancho San Augustin and Rancho Zayante, totaling just over 7000 acres. He became a leading citizen and was elected as the first American Alcalde of Santa Cruz in 1845, which gave him the combined role of judge and mayor.

Joseph Ladd Majors, his son Joseph Joaquin Majors, and two of his grandsons, owned cattle and dairy ranches on the North Coast, although it is unclear which of the family member(s) the beach, the creek and the settlement were originally named for.

Shortly after arriving in Santa Cruz in September 1968, I headed north on Highway 1 one Saturday afternoon looking for access to the shoreline for places I might take students for geology field trips. I turned off at the first road I encountered that seemed to head towards the shoreline. To my surprise, I soon ran into a gate and a tollbooth.

After inquiring whether there would be a problem bringing students down to look at the cliffs and beach, the caretaker told me that this was a clothing optional beach and it might not be the best place for a field trip. For 40 years the road to the beach was marked with a Red, White and Blue mailbox right off Highway 1 and was well known destination for campers and day trippers. The popular beach closed in 2007, however.

Heading south, the next easy public access is usually marked by lots of cars parked on the ocean side of Highway 1, about 4 miles north of the city limits and Western Drive. Known as Four Mile Beach, this point break has been a favorite surfing spot for years, at least since I arrived in Santa Cruz. My experience has been that while it's a bit of a walk from the parking area, its much less crowded than many other spots in town.

The formal historic name for the beach was Baldwin Creek Beach, named after the creek that empties into the ocean here.

Although there are two early Santa Cruz pioneers with the name Baldwin, who were unrelated, Donald Clark in his book believes that the creek and beach were named after Alfred Baldwin.

    Alfred migrated to California in 1846 from New York, and settled in Santa Cruz in 1847. He served under Captain John C. Fremont, spent some time looking for gold in the Sierras, but returned to Santa Cruz where he was a shoemaker and also owned clothing stores. Baldwin owned a farm on the North Coast in the vicinity of Baldwin Creek. The confusion regarding whom the creek was actually named for arose when Alfred Baldwin sold the farm to Levi Baldwin. What is unclear, at least to me, is whether these early landowners named the creeks, or whether the names became attached more informally by the local population over time.

    One of the remarkable parts of the lives of both Joseph Majors and Alfred Baldwin is that they, like a number of other rugged and adventurous individuals, left what perhaps were comfortable lives and homes in the east, and struck out for California with little knowledge of what to expect along the way or what they might find when they got here.

    Gold hadn't been discovered yet, so that wasn't the attraction. These guys were committed and it took months across some pretty unfriendly terrain to get here in the 1840s. Lots of folks didn't make it. Today it's a piece of cake; buy a ticket and fly out in several hours. Rent a car at the airport and you can be in Santa Cruz on the same day you left New York. All you need is $500 and if you don't like it, you can get a red-eye, fly back and be home the next morning.

Article 19

Wilder Ranch Still Wild

We are fortunate today to have Wilder Ranch State Park and all of its uncrowded beaches, coastline and terraces, as well as its 34 miles of trails, literally in our back yard. In the late 1960s we almost lost those 4000 acres and that coastline to a large development.

In 1871, Deloss Wilder, in partnership with L. K. Baldwin (of Baldwin Creek fame), bought 4000 acres with two and a half miles of ocean frontage from Moses Meder, another early Santa Cruz pioneer. Wilder, like most of the area's early settlers, came from the east, originally Connecticut. He headed west to seek his fortune in the California Gold Rush, but after six years of digging, and not finding the fortune he had hoped for, Wilder moved west to Marin County in 1859 and started a dairy and chicken ranch. In 1871 he moved south to Santa Cruz, and along with Baldwin, built up the largest and most productive dairy in the county.

In 1885, Baldwin and Wilder dissolved their partnership and the ranch, with Wilder taking the parcel closest to the city. If you bike to Wilder Ranch today, and turn down the first entry road to the ranch buildings from the bike trail, you will be riding downhill on a stretch of pavement that was the original Highway 1. The old highway passes right in front of the old ranch buildings and farm houses.

Deloss Wilder and his two sons ran a herd of 350 dairy cows, three milk routes through the city and churned a ton of butter every day. The Wilders also developed a successful waterpower system using the water pressure from Wilder Creek through a device known as a Pelton Wheel.

The Pelton wheel was invented in 1870 and is very efficient at using the force of moving water that has a low flow volume but a large elevation difference, or hydraulic head. This was ideal for using the small flow of water from Wilder Creek that was at a high enough elevation upstream from the ranch to generate power.

The ranch used this mechanical power to run a number of tools in the barns on the ranch, such as saws, lathes, and water pumps that are still there today, as well as a small electrical generator that produced what was reportedly the first electrical lighting system in Santa Cruz.

Wilder's sons, Deloss Jr. and Melvin, operated the dairy until about 1937 when the ranch turned to cattle raising and also leased out the coastal terrace land for artichokes and Brussels sprouts.

The economics of ranching and farming changed along the coast in the late 1960s, however. Although I am not sure who made the decision, property values at that time were assessed and taxed based on their development potential, not their value as farmland.

The dwindling income from farming and the increased property taxes led the Wilder family to sell to a land development firm in 1968. That transaction changed ownership of the land for the first time in 97 years, and nearly altered the entire character of this special piece of coastal landscape.

A Canadian development firm bought Wilder Ranch, and by 1972 had developed a plan based on a "new town" concept, which would have ultimately constructed seven "villages" that would have housed 35,000 people.

The next time you bike or hike on Wilder Ranch, or drive by, try to imagine 35,000 additional people filling up those 4000 acres of special places where you can find solitude and nature preserved today.

It's equally difficult today to think about anyone even proposing such a project. Yet it was very real at that time and immediately sparked a controversy about how that coastal land should be used, as well as the impacts of nearly doubling the size of the city.

A small group of people initiated an effort to oppose the development of the property and ultimately, the conservationists and environmentalists, who were in relatively short supply 42 years ago, won and the state bought the property in 1974 for a new state park. Thanks to the early efforts of those radical environmentalists, we can all enjoy the land, the biking and hiking trails, and the coastline and beaches today.

Article 20

Beach Trash or Beach Glass?

Glass Beach, Fort Bragg, California.
(© 2006 D. Shrestha Ross)

Beach glass has become somewhat of a treasure in recent years, although it certainly didn't start out that way. Fort Bragg, an old lumber town on the Mendocino coast, has a well-known beach on the north end of town, appropriately known as Glass Beach. The town now even has a Glass Beach Inn.

This beach was the city dump for decades, well before the days of recycling and home pickup of trash. Residents just dumped their cans, bottles, broken plates and bowls, old refrigerators and stoves, or whatever else they had lying around the backyard, over the low bluff onto the beach. This was pretty standard way of getting rid of trash in the 19th and early 20th centuries. Fires were set from time to time to burn what was combustible and most of the metal gradually rusted away.

But the resistant pottery and glass has been gradually broken down, smoothed and polished by the waves, leaving a sparkly beach with green, brown, clear and occasionally blue beach glass. Dumping on the beach was finally terminated in 1967.

While attempts were made to remove the material to "clean up the beach", this unique beach was eventually preserved as part of MacKerricher State Park and is now enjoyed by thousands of visitors every year. While removing glass isn't officially permitted, it's probably safe to say that every visitor walks away with some treasures in their pocket. You might want to be very careful, however, if you are trying to cart off a wheelbarrow or bucket load.

Glass Beach is hardly alone, and we have our own glass beaches; they just haven't been recognized and named. As recently as the late 1960s, the residents of Davenport dumped their trash and old appliances down the railroad embankment overlooking Davenport Beach. From time to time the debris was set on fire, but lots of stuff didn't burn very well. Eventually the practice was halted, and the waste was covered over with rock and dirt. Very little of the debris, however, ever made it to the beach to be reworked and polished by waves. It still lies buried on the side of the railroad embankment.

There is some beach glass to be found at the north end of Davenport Beach if one doesn't mind digging around a bit with other treasure seekers, but it has quite a different source. A small glass blowing business has operated for years alongside San Vicente Creek on the opposite side of Highway 1. The creek flows through a hand-dug tunnel under the railroad and Highway 1 embankment and has carried small pieces of colorful glass onto the beach where it has been smoothed and polished. The place is no longer a secret, however, and people are often seen on the beach at the mouth of the creek with shovels and screens, searching for buried treasure.

The old Wilder Ranch had their own beach dump as well. Up until about 1970, the narrow cove north of Fern Grotto, which formerly had an arch on its seaward end that was often photographed with horse drawn buggies perched on top of it, was the trash dump for the ranch. Whatever was broken, worn out, or had no useful purpose, was dumped down the embankment onto the beach.

The green, brown, and clear glass was reworked by the waves, and during the winter months when much of the sand has been moved offshore, leaving the coarser material behind, Wilder beach glass can be found on Fern Grotto Beach. Glass isn't rounded and polished overnight, however, so anything you find has a history going back at least 50 years, and probably much longer.

Today, leaving glass on the beach, or anything else for that matter, is punishable by death, but it was common practice 50 or 100 years ago. We didn't have curbside recycling to make it easy back then, and few people visited these beaches. But today, if you have sharp eyes and get to the beach before anyone else at a winter low tide, you just might find the polished remnants of one of your grandma's old plates or your grandpa's old beer bottles.

PART II

GEOLOGY & FIELD TRIPS

Article 1

Peeling Back the Layers

Sandy Lydon: "You call this a science?".
(© 2012 D. Shrestha Ross)

Twenty-five years ago, I was on a raft trip down the Grand Canyon of the Colorado with my good friend and local historian, Sandy Lydon. The trip was a non-commercial adventure, run by another good friend and former student, local geologist Jerry Weber. Private trips get non-prime time, which in this case meant departure on the 1st of April. One of the unplanned consequences of leaving that early in the season is that it was snowing during our departure from Lee's Ferry, which was a bit disconcerting.

Interesting things happen when geologists, who tend to think in terms of millions of years, hang out with historians, who tend to think of a few to perhaps several hundred years. The Grand Canyon provides an exceptional exposure of about two billion years of Earth history that are recorded in the walls of the canyon. Much of the limestone, sandstone and shale exposed in the 5000-foot deep canyon was deposited in an ancient ocean many hundreds of millions of years ago. This region of the Earth wasn't stable throughout this entire period of time, however, and as a result, after sediments filled an ancient ocean, the crust was uplifted or raised by tectonic forces within the Earth, and erosion took place. The same thing is happening in the Santa Cruz Mountains today every time it rains. The marine sandstones and shales you see exposed in road-cuts along Highway 9 or 17 as you commute to work, are gradually being eroded and carried down hill, through the creeks and rivers, to be deposited offshore on the seafloor of Monterey Bay.

While the Earth is about 4.6 billion years old, there is no known place on Earth where sedimentation or deposition has been continuous throughout geologic time and where we have the entire book of Earth history recorded. In fact, to confuse things further, there aren't even rocks on Earth that are 4.6 billion years old. So we use the age of moon rocks to give us a better estimate of the age of the Earth.

Tectonic uplift or subsidence has periodically interrupted the relative stability of the Earth's crust and thereby confused and removed some of the history recorded in the rocks. During the formation of a continent, for example, large sections of crust are raised out of the shallow seas in which sedimentary rocks were deposited, subjected to erosion, and then often subside again below sea level where deposition of sediment begins again.

Activity of this sort produces a buried erosional surface, with older rocks below that are covered with younger sediments. These erosion surfaces, called unconformities, separate rocks that may be of vastly different ages depending upon how much time elapsed between the erosion of the older rocks and the subsequent deposition of younger sediments. We can see this process exposed locally in the seacliffs along West Cliff Drive. The mudstone and sandstone of the Purisima Formation exposed in the lower portion of the cliffs is about 3 to 5 million years old, but is capped or overlain by 5 to 10 feet of sands and gravels that are only 100,000 years old.

When geologist and one-armed Civil War veteran Major John Wesley Powell first ventured down the Grand Canyon with 10 men and 4 boats in 1869, one of the important observations he made was of what he named the Great Unconformity. He discovered a section in the canyon walls where there was a 600 million year gap between the sedimentary rocks below and above the unconformity. Sandy Lydon, a responsible historian, who feels the need to account for every year of the historical record, nearly fell out of the boat when this section of lost geologic time was explained. "How can you geologists just leave out 600 million years of history? And you call this a science?"

ARTICLE 2

Living on an Ancient Seafloor

Even if you never go to the beach, chances are if you drive anywhere in Santa Cruz you are traveling across an ancient beach. Much of the city has been built on an uplifted marine terrace, an old chunk of seafloor now covered with houses, stores, streets, and if you head north, Brussels sprouts. If you start peddling your bike at West Cliff, working your way up Bay Street, you'll notice it's a gentle climb all the way to Escalona Drive. While it's easy driving up that hill, on a bike you notice immediately that you are climbing. Escalona forms the inner edge of the first marine terrace and is the base of an ancient sea cliff. If you are further west and cross Mission Street at Western Drive, the climb is even steeper. It's obvious even in a car that you are climbing a cliff, an old sea cliff.

Dating of fossils recovered from the terrace surface where it is exposed along West Cliff indicate that this wide, nearly flat bench was formed about 100,000 years ago. Terraces have made development of many of California's coastal communities possible: Mendocino, Half Moon Bay, Santa Cruz and Capitola, for example. Where we don't have terraces, Big Sur for example, there is almost no development, and even building Highway 1 was challenging because of the steep slopes.

Why are these flat terraces so common along California's coast? Creating and then preserving marine terraces requires several things: a slow but continuous uplift of the coastline, an oscillating sea level, and bedrock that is weak enough to be eroded by waves. For millions of years the edge of what is now California was a place where two massive crustal plates collided. While the offshore oceanic plate was forced down under the advancing North American plate, this collision also led to gradual uplift of the coast of ancient California, helping to create the Coast Ranges. Some of this slow uplift continues today.

The second requirement is that sea level rises and falls over time. We know that the Earth's climate has changed constantly over time scales of thousands of years because of the amount of heat we get from the sun. This will be described in a later column so you'll have to accept it for now. Simply put, the closer we are to the sun, the warmer the Earth gets; the farther we are away, the cooler it gets. When the Earth is warm, as it is today, the ice sheets and glaciers begin to melt and sea level rises in response. At times of high sea level, the wave are doing their job, washing sand back and forth across the intertidal zone, grinding down the bedrock and forming a flat rocky terrace, visible at low tide today. When the climate cools again, sea level drops, and if the land is rising, we will elevate and preserve a flat marine terrace. Variations in local conditions have lead to the preservation of up to five terraces in the Santa Cruz area, with Wilder Ranch being a great place to see them on a bike.

Article 3

Our Flooded Edge

Several weeks ago I wrote how most of us live, drive, and bike or walk on an ancient seafloor every day in Santa Cruz. I tried to explain how the interaction of long-term fluctuations in sea level, wave erosion in the surf zone, and gradual coastal uplift over perhaps 500,000 years had produced a set of flat, elevated marine terraces that we call home. The part of this story that may be the most difficult to grasp, but which is really important to understand as we think about climate change today, is that of sea level rising and falling.

If we look any of those very cool maps of the seafloor of Monterey Bay, or the seafloor off virtually any coastline in the world, we will see a nearly flat, relatively smooth submerged terrace - the continental shelf. This shelf extends only a few miles offshore in Big Sur, about 8 to 10 miles off Santa Cruz, and nearly 25 miles off of San Francisco. In fact, the Farallon Islands sit on the continental shelf. The outer edge of the continental shelf is usually about 350-400 feet deep, and at that point the ocean bottom drops off more steeply to the deep sea floor some 10,000-12,000 feet below.

Studies of the continental shelf around the world over the past 75 years have revealed that this flat feature was created by sea level rising and falling and waves washing back and forth across the edge of the continent many times. Each time this happened, the coastal cliffs were eroded back farther, and the seafloor was gradually smoothed as breaking waves eroded off the high points and sand and mud filled in the depressions. Old beach sands, intertidal and shallow water fossils, and even peat beds, which form in fresh water bogs, have been recovered from the submerged continental shelf hundreds of feet below present sea level, providing evidence that this area has been swept over repeatedly by the ocean, leaving behind a preserved record of that history of alternating exposure and inundation.

Lowering sea level 400 feet exposed a land bridge across the Bering Sea, which allowed early humans to migrate from Siberia to North America perhaps 20,000 years ago, and then gradually work their way south, and slowly populate the Americas. At the same time, the English Channel between England and France was dry land and allowed for migration back and forth. And we could have walked from San Francisco out to the Farallons without even getting our feet wet.

There is a lot of real estate offshore that is gone forever, at least for the next several thousand years, perhaps longer. While we lease the continental shelves off southern California, as well as Louisiana and Texas, for oil drilling and extraction, it isn't going to provide any new home sites in the near future.

A logical question you might be asking at this point is where did all of those 10,000,000 cubic miles of ocean water go that were evaporated from the oceans in order to lower sea level 400 feet? That story will have to wait for another column.

Article 4

Sand Dunes-Blowing in the Wind

Sand Dunes along the southern Monterey Bay shoreline.
(© 2012 D. Shrestha Ross)

Even with hundreds of dams restricting river flows, California's streams still provide over 10 million cubic yards of sand annually to the state's beaches. It would seem that the continued addition of all of that sand to the coast would lead to wider and wider beaches. Once arriving at the shoreline, however, sand tends to get moved down coast due to the dominant waves from the northwest along most of the California coast. Since the beaches aren't getting wider, except where we have built barriers to sand transport such as breakwaters, jetties or groins, where does all that sand go?

This river of sand or littoral drift may carry individual sand grains 50 to 100 miles or more along the shoreline in some littoral cells, where they are eventually lost either onshore into dunes, or offshore into submarine canyons. Sand dunes occur inland from some beaches along the California coast and they can act as sinks where beach sand is permanently lost from the shoreline.

Huge volumes of beach sand historically were blown inland at Ocean Beach in San Francisco, along the central and southern Monterey Bay shoreline, at Pismo Beach and Nipomo Dunes, along the Oxnard Plain, and from Santa Monica to El Segundo. Some of these dune areas are enormous and account for a lot of lost beach sand. In the 1800's, dunes extended from Ocean Beach near Golden Gate Park clear across the San Francisco peninsula to the bay. Driving to Monterey you pass through the middle of a huge dune field covering over 50 square miles that extends from La Selva Beach to Monterey, much of it lying under the old Fort Ord military base.

In order for any significant area of sand dunes to form, we need to have a large supply of fine-grained sand, a wide expanse of beach such that an area of dry sand exists for at least part of the year, low relief topography landward of the shoreline that provides a place for the sand to migrate and accumulate, and equally important, a dominant onshore wind direction. As a beach widens and the area of dry sand on the back beach expands, a persistent onshore wind can begin to move the fine-grained sand off the beach and inland.

Most of California's large fields of sand dunes formed under different conditions that we experience today, however. Twenty thousand years ago during the last Ice Age, sea level was about 350-400 feet lower than at the present and the shoreline was considerably seaward of where it is today. In central and southern Monterey Bay, the shoreline was about 8 to 10 miles to the west and the Salinas and Pajaro rivers crossed the now exposed continental shelf. These rivers drained large inland areas and the sand they transported and then deposited across this broad exposed shelf was subsequently blown inland by persistent onshore winds to form the vast dune fields that now underlie Marina, Fort Ord, Seaside and Sand City.

Sea level subsequently rose as the Earth warmed and ice sheets and glaciers melted; the shoreline retreated back across the shelf, and the dunes began to erode. Today the dunes along the shoreline of southern Monterey Bay are retreating at two to over six feet every year on average.

Article 5

Natural Curves Along the Shoreline

The hook-shaped shoreline of northern Monterey Bay
showing beach cusps, 1956
(photo: US Department of Agriculture, courtesy: UC Santa Cruz Map Library)

A high altitude aerial photograph of the California coast reveals some perfectly curved sections of shoreline. Half Moon Bay was named after its smooth curved shape. Bodega Bay, Drakes Bay, Stinson Beach, San Pedro Bay in its pre-breakwater configuration, and the Silver Strand in San Diego are other good examples of shorelines, which still have or had a nearly perfect hooked shape, uncoiling or unwinding from north to south.

Monterey Bay is unique in having smooth curved beaches at both ends. The northern end actually has two hook-shaped sections; one begins at Cowell and extends south and east towards Blacks Point, now interrupted by the jetties at the harbor. The second begins along Depot Hill in Capitola and gradually unwinds with a smooth curve extending all the way to Moss Landing. At the

Part II: Coastal Geology & Field Trips

southern end of the bay, the irregular granitic coastline of the Monterey peninsula changes at the Monterey breakwater into a smooth sandy beach that curves gently upcoast all the way to Moss Landing. This curve is interrupted only once, by the bulge of sand formed by the delta at the mouth of the Salinas River.

Each of these bays begins with a tight curve downcoast of a rocky point, and then gradually uncoils proceeding alongshore, just like the shell of an abalone or some other mollusk as it grows. These smooth uncoiling shorelines owe their origin to the process of wave refraction, or the bending of wave fronts as they approach the coastline. They form where rocks occur along the coast that differ in their hardness or resistance to wave attack. Where a hard rocky headland occurs, and there is a dominant direction of wave approach, the stage is set for the formation of one of these uniquely shaped features.

Half Moon Bay is a good example, where Pillar Point stands out as a headland rising some 150 feet above sea level. Under natural conditions, the dominant waves approached from the northwest and then wrapped around Pillar Point and attacked the low-lying and softer sedimentary rocks immediately downcoast. The bending or refraction of the waves as they wrapped around the point spread out or diffused the wave energy in a very regular pattern. The refracted waves gradually eroded the weaker rocks downcoast and behind the point and, over time, the shoreline or beach gradually developed a gently curved or uncoiling shape that mirrored the wave fronts of the bending waves.

When the Half Moon Bay breakwater was built in 1959, however, the wave energy that was formerly dissipated along the shoreline of this smooth bay was now concentrated at the south end of the new breakwater. The low weak bluffs adjacent to Highway One here are now eroding at about five feet per year and threatening the highway. The breakwater disrupted the natural pattern of breaking waves. In contrast, the construction of the west jetty at the Santa Cruz harbor in 1963 acted as a dam for downcoast sand transport. Seabright Beach, which was originally quite narrow, has been reshaped with the curved shoreline in this instance growing seaward, widening the beach by hundreds of feet.

A series of very symmetrical scallops are visible along the beach face in the aerial photograph accompanying this article. These semi-circular features are called beach cusps and are very common between Cowell Beach and the harbor. You can observe them as you look down on the beach from the bluff top or notice them as you walk along the beach and realize you are going up and down across a gently rolling surface. These features can be uniformly spaced, 50 to 150 feet apart, and are typically aligned with small, evenly spaced rip currents that flow seaward from the individual embayments. We don't completely understand why these features form, although there are two different competing theories

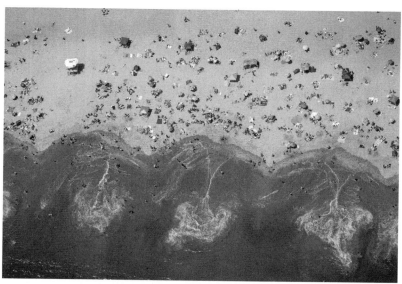

Beach Cusps, Main Beach, Santa Cruz.
(© 2010 Jeremy Lezin)

Part II: Coastal Geology & Field Trips

ARTICLE 6

Submarine Canyons
Sediment Highways to the Deep Sea

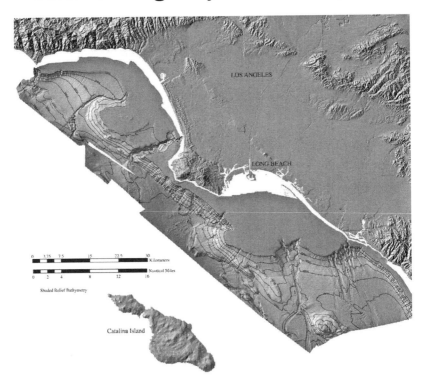

Onshore topography and offshore bathymetry of Southern California.
(courtesy: the US Geological Survey)

Along most of California's coastline, littoral drift is carrying the beach sand southward. I asked in an earlier story, where is all of this sand going, and why aren't the beaches growing wider and wider as you move down coast? There are only a few places all of that sand can eventually go: blown inland as dunes, removed by mining, or transported offshore.

The greatest loss of sand from California's beaches is invisible to us and occurs down the many submarine canyons that wind their way across the seafloor just offshore. Where these canyons extend close to shore, they intercept the sand moving along the coast and funnel it away from the beach down to the deep-sea floor.

The canyons off southern California have been recognized for over 75 years as the ultimate sinks for most of that area's beach sand. The Mugu Submarine Canyon, south of Ventura, siphons off over a million cubic yards of sand each year, enough to build a beach 100 feet wide, 10 feet deep, and 5 miles long! A number of other offshore canyons remove sand from the shoreline as well; the Zuma, Redondo, and Newport submarine canyons play similar roles.

Monterey Submarine Canyon, which bisects Monterey Bay and extends almost to the beach at Moss Landing, is one of the world's largest-over 6,000 feet deep-and big enough to swallow the Grand Canyon of the Colorado River, but completely invisible to us standing on the beach! Every year, virtually all of the nearly 300,000 cubic yards of sand that is transported down the coast from Santa Cruz, some from as far north as Half Moon Bay, as well as the sand carried northward along the shoreline from the Salinas River, is carried offshore into deep water by this vast underwater conveyor belt.

Observations by SCUBA divers and submersibles reveal that in the steeper canyon heads this sand flows down slope, grain by grain, the same process that occurs when you start digging into a sand pile or dune. Transport farther offshore, however, where the slopes are less, is achieved by underwater sand and mudflows known as turbidity currents. Turbidity currents are large masses of sediment that are driven by their greater density relative to seawater, and are capable of flowing many miles down submarine canyons.

Millions of cubic yards are lost from California's beaches into submarine canyons each year, and this is why the beaches don't get any wider. Ultimately, these former beach sands are deposited far offshore in deep water. With their final resting places at depths of 10,000 to 15,000 feet below sea level, these sands are in effect, gone. Concerns over the losses of beach sand due to dams on coastal streams have led to proposals to dam these offshore canyons as a way to trap this sand so that it can be pumped back onto the beaches. Dredging sand up from deep water would be a very expensive process and it would be far easier, and would consume much less energy, to halt or trap the flow of sand at the shoreline before it moves offshore into a canyon head.

Article 7

Grand Canyons on the Seafloor

The presence of canyons on the seafloor, similar to those on land, came as a surprise to early oceanographers who first noticed them on their depth recorders, slicing into the continental shelf and slope off of the mid-Atlantic coast over a century ago. As active scientific exploration of California's offshore area began in the early 1900s, a number of submarine canyons were soon discovered here also. Those immediately offshore of the Scripps Institution of Oceanography at La Jolla, appropriately named Scripps Canyon and La Jolla Canyon, were among the first charted and studied in detail. Before long, canyons were discovered scattered along the entire length of California's coast.

Most of the largest submarine canyons were given names, usually based on their proximity to some nearby community or landform, so we ended up with Coronado Canyon, Newport Canyon, San Pedro Canyon, Redondo Canyon, Santa Monica Canyon, Dume Canyon, Mugu and Hueneme canyons along the southern California coast. Carmel, Monterey and Soquel canyons are the most prominent submarine canyons along the central coast.

Submarine canyons are similar in many ways to river canyons or drainage systems on land. They are dominantly erosional features that are carved into the continental shelf and slope, often exposing bedrock in their walls. They typically have windy or sinuous courses and may have a number of tributaries. All submarine canyons extend from the outer edge of the continental shelf, down the continental slope to the deep sea floor. While some of California's canyons extend no closer to shore than the edge of the shelf, others extend completely across the shelf nearly to the beach. Most of the prominent southern California canyons, as well as Carmel and Monterey canyons, extend into shallow water near river mouths, which gave some clue as to the origins of these massive and mysterious sea floor features.

Following the initial discovery of submarine canyons off the Atlantic coast, and for many years thereafter, there were two competing views on the origins of these little understood seafloor features. Because they seemed to resemble river valleys or canyons on land, and because many of the submarine canyons first surveyed were located offshore from river mouths, some scientists believed that these undersea features must have been eroded by the adjacent rivers when sea level was lower during past Ice Ages. We knew that the continental shelves were exposed during those glacial periods, so the belief was that rivers simply continued cutting their channels across the now exposed sea floor. This wasn't a problem for the sections of the canyons that crossed the continental shelf, which only extended to depths of 350 to 400 feet, but presented some serious difficulties for the portions that extended down the continental slope to the deep sea floor, some 10,000 feet deeper.

Relying on rivers to cut the canyons to depths of 10,000 feet or more would have required draining most of the water from the world's oceans. To say the least, this is a challenging undertaking. Over time, however, the advocates for the river erosion hypothesis for submarine canyons realized that emptying the oceans simply wasn't geologically possible. There just wasn't any place to put the roughly 325 million cubic miles of seawater. So we needed to come up with a better idea for their formation.

Article 8

Submarine Canyons - Going Deeper

Tubidites, Point Lobos State Reserve.
(© 2008 D. Shrestha Ross)

While the early marine geologists who believed that submarine canyons might have been eroded by rivers and then were later submerged had to reconsider their ideas, there was an alternative view gaining support, that underwater processes had cut the canyons, or that they had a submarine origin. The major shortcoming with this new idea was that there wasn't any mechanism that had been observed 50 or 60 years ago that seemed capable of cutting canyons thousands of feet deep on the seafloor, and in the case of portions of Monterey and Carmel canyons, apparently through solid granite! Several subsequent discoveries, however, began to help answer the question of how these Grand Canyons of the seafloor could have formed underwater.

Layers of sand were discovered in the first sediment cores that were recovered from the floors of these underwater canyons. Marine geologists for years have dropped corers, long steel pipes with heavy weights attached, into the seafloor from ships to extract cores of sediment. The sediment layers provided a history book or record of what processes had taken place over time at that particular location in the ocean. When the sand layers from submarine canyons were first recovered, we recognized that the layers were graded, or the sediment was coarsest at the bottom and finest at the top. This graded bedding can be created in a jar of mixed sediment (gravel, sand, silt and clay, for example) and water, if we shake it up and let the sediments settle out. The gravel settles fastest and ends up at the bottom of the jar, followed by the coarsest sand, then finer sand, and finally silt, and hours later, the clay. So the graded sediments recovered from the floors of submarine canyons indicated that these sediments had settled out of suspension.

Additionally, even though many of these sediment cores were recovered from depths of hundreds or thousands of feet below sea level, and many miles offshore, they contained shallow water fossils, indicating that the sediments originated in shallow water. The concept that emerged from these observations was that underwater flows of muddy and sandy sediments, which were given the name turbidity currents, could be generated in shallow water at the heads of these submarine canyons, perhaps from earthquakes, large storm waves, or river floods. These underwater avalanches would flow down slope along the seafloor, driven by gravity because the muddy sediments were denser than seawater. Sand is abrasive, which is why we use sandpaper to smooth wood, and would enable these turbidity currents to progressively cut canyons across the continental shelf and slope over thousands of years.

Geologists working on land have recognized preserved turbidity current deposits, called turbidites, exposed in outcrops of old sea floor deposits in mountain ranges all over the world. In fact, take a drive down to Point Lobos State Reserve some weekend, just south of Carmel, and you can walk through a spectacular coastal display of 50 to 60 million year turbidites exposed along the entire south coast of the reserve. All evidence suggests that these thick graded layers of gravel and sand were deposited in an ancient submarine canyon by turbidity currents in an environment much like Monterey Submarine Canyon. If you prefer to drive north, there is another exposure of identical turbidity current deposits exposed in the coastal cliffs from Pigeon Point northward. There is more to this story.

Article 9

Why Submarine Canyons?

In 1929, there was a large earthquake in the Grand Banks area off Nova Scotia that provided the evidence need to confirm the existence, and ultimately the importance of turbidity currents to submarine canyon formation. Hard to believe, but in 1929 we had no cell phones or satellites to relay phone calls, so a number of very long underwater cables had been strung along the seafloor to connect phone lines in North America with Europe. In the minutes immediately following the earthquake there was a well-documented cut-off in phone connections as these cables progressively failed proceeding down slope away from the earthquake's epicenter. It appeared to the scientists who studied these cable failures that a flow of sediments along the sea floor, or a turbidity current, had been generated by the earthquake and had sequentially broken the telephone cables as the sediment flow progressed down slope to the deep sea. The confirming evidence was obtained when oceanographic vessels subsequently cored the sea floor in the area and found graded sands at the surface as evidence of the passage of a turbidity current.

Putting together these and other observations has produced a coherent picture of how submarine canyons have formed. It turns out that both terrestrial and underwater processes are involved. We need to have at least a portion of the continental shelf exposed to initiate the cutting of a submarine canyon, which happens where a river discharges. As sea level drops during a glacial period, the shoreline retreats but the river continues to erode its channel across the shelf in order to reach the ocean. At least half of all of the submarine canyons studied lie directly offshore from rivers, providing good evidence for the role of rivers in the initiation of many submarine canyons.

With each rise in sea level, the river retreats back across the shelf where it discharges at the new shoreline. With the next drop in sea level, the river will again work its way across the shelf, further deepening its channel. The river's sediment load will now be discharged at the outer edge of the shelf where it will accumulate until it becomes unstable, and then, either from an earthquake or some other disturbance, the sediment will cascade down the steeper continental slope as a turbidity current. Abrasion by the coarser sand and gravel will gradually deepen and widen the canyon, while the finer silt and clay may be deposited along the banks of the canyon as natural levees, similar to a river in flood stage. Thus a submarine canyon can grow by both eroding its channel downward, and by constructing natural levees upward.

An essential process maintaining those submarine canyons along the California coast today is the active transport of littoral sand from the beaches into the canyon heads. Most of California's littoral cells terminate in submarine canyons. As waves from the northwest drive littoral drift southward, the sand at the end of each cell is often carried offshore into the head of a submarine canyon. Hundreds of thousands of cubic yards of beach sand is funneled down each these canyon heads every year, to ultimately be carried by turbidity currents down to the deep sea floor 10,000 to 12,000 feet below.

Part II: Coastal Geology & Field Trips

ARTICLE 10

Why Monterey Submarine Canyon?

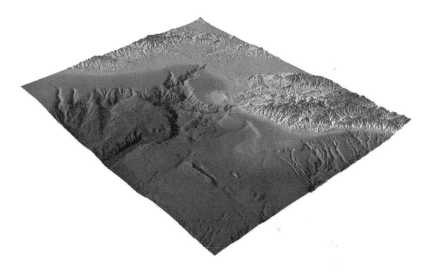

Relief of the Monterey Bay Region both on land and the seafloor.
(courtesy: US Geological Survey)

At the risk of beating submarine canyons to death, I'm going to try to wrap up this discussion with some final thoughts on why we have one of the world's largest submarine canyons in our backyard. Monterey Submarine Canyon has been known for over a century, and as with other offshore drainage systems, there has been considerable speculation over the years as to why we have this huge chasm cutting across the seafloor. Most submarine canyons align with river systems, but Elkhorn Slough hardly provides an adequate onshore source for such a massive feature. We do know that prior to 1910 the Salinas River discharged six miles north of its present mouth into Elkhorn Slough, closer to the head of Monterey

Submarine Canyons. But even the Salinas River, however, is not of the scale of a river we would expect for an offshore feature as large as the Grand Canyon.

Over 50 years ago, two geologists discovered the presence of a deep buried inland canyon from oil company drill holes beneath the Santa Cruz Mountains. This combined with other geological and geophysical observations strongly suggested that this canyon was eroded by an ancient river drainage system that played a critical role in the initial formation of the Monterey Submarine Canyon. The buried canyon, named Pajaro Gorge by some, is believed to be the route that the drainage from California's vast Central Valley followed to the ocean for million of years. The large Sacramento and San Joaquin river systems, which drain the Sierra Nevada and most of the interior of California, have only discharged through the Golden Gate for about the last 600,000 years. Prior to that time, we believe they worked their way to the ocean much farther to the south, with Pajaro Gorge being their path for millions of years.

Further complicating the history of this interconnected terrestrial and submarine drainage system is the San Andreas Fault, which cuts through the Santa Cruz Mountains just behind Watsonville. All of us living in Santa Cruz County are heading northwest towards Alaska at about two inches per year as the Pacific Plate slowly grinds its way past the North American Plate to the east. Every million years or so, we move another 25 miles away from the location on the opposite side of the fault in our northwesterly journey, ultimately to become a Madagascar of the Pacific. Monterey Submarine Canyon is moving along with us, which over time has taken it farther and farther from the original inland drainage system that initiated this offshore canyon. Ten million years ago, ancestral Monterey Bay and the canyon's sediment sources were 250 miles to the south. Cores taken from the deep-sea fan that has formed where Monterey Submarine Canyon ultimately deposits its sediment reflect the changes in the geology of the source areas where the sediments have come from over time.

What we see today is an old submarine canyon, formed millions of year ago under very different geographic conditions that those we see now. The sand moving into the canyon head at present comes from littoral drift from both northern and southern Monterey Bay beaches, rather than through a deep gorge from the Central Valley as it did in the past.

Article 11

Ancient Mud, Diatoms, and Whales

Whale vertebrae
preserved in the Purisima Formation, Depot Hill, Capitola
(© 2008 Gary Griggs)

Westsiders and eastsiders in Santa Cruz each have their own world-class surfing spots; they also each have distinct bedrock and coastlines. The rocks underlying West Cliff from Almar Avenue north to beyond Davenport are mudstone, the Santa Cruz

Mudstone to be precise. Although the mudstone does erode over time, this stuff is actually pretty hard because of its high silica content; in some places it's almost like porcelain and it puts up a good fight against the incessant wave attack. Most of the silica came from the shells of the billions of ancient diatoms that populated the sea that covered the Santa Cruz area 7 to 9 million years ago when the mud was deposited.

From Almar Avenue southeast to Rio Del Mar, the cliffs consist of the Purisima Formation, which is younger by several million years and which is a mixture of sandstone, siltstone and mudstone. Most beach goers in Santa Cruz County have a chunk of Purisima stuffed with fossil clam shells somewhere on their front porch, window sill or book shelf. If you walk along the shoreline below Depot Hill in Capitola on a low tide, and make sure it's a low tide or you won't get very far, keeping your eye on the overhanging cliff, you can see huge slabs of Purisima that have fallen to the beach. Many of these are molluscan graveyards. If you look carefully, you can also almost always find a few fossilized whale bones embedded in the rock as well. In contrast to the Santa Cruz Mudstone, the Purisima is much softer and weaker, and is also extensively jointed or fractured so yields more readily to wave attack.

For ten miles or so of the coastline of northern Santa Cruz County, sedimentary intrusions, or sandstone dikes and sills, have been well preserved within the mudstone seacliffs. Most of us are more familiar with volcanic intrusions, where hot magma under pressure works its way upward towards the Earth's surface and leaves behind dikes and sills of hardened lava. Somewhat surprisingly, there are also sedimentary or cold intrusions. These seemed to have been created when deeper, fluid saturated sediment was liquefied like quicksand, perhaps by a long ago earthquake, and forced upward like toothpaste squeezed out of a tube.

From the beaches of Wilder Ranch to beyond Davenport, a bizarre variety of these unique features are exposed along the coastline. Near the former railroad stop and settlement of Majors on a short stretch of the old Coast Road (about 5.5 miles north of the city limits), they can actually be seen from Highway One in road-cuts without even getting out of your car. In other cases, a short walk to one of the nearby pocket beaches will reveal some landforms not visible in many other places in the world.

While their origins aren't completely clear, an early geologist working his way up the coast first recognized these intrusions a century ago in the cliffs south of Davenport. Yellow Bank Beach is named after the yellow or orange sandstone exposed in the cliffs, believed to be the largest exposures of this phenomenon anywhere on Earth! You can even see a few remains of the old Yellow Bank Dairy on the inside of Highway 1, named after the coloration in the cliff (which actually looks more orange to me).

Some outcrops of the sandstone intrusions are oil-bearing and form natural asphalt deposits. These were quarried at several sites a mile or two inland in Bonny Doon from about 1888 to the 1940's. This asphalt was reportedly transported to San Francisco in the 1890's to pave the city streets.

Part II: Coastal Geology & Field Trips 97

Article 12

A Walk to Remember

Satellite image of Monterey Bay, California, 1998.
(courtesy: National Aeronautics and Space Association)

Have you ever looked at that wonderful high altitude, composite color photograph of Monterey Bay and wondered what it might be like to walk completely around the bay? Can you even walk all the way around it? Eighteen years ago, Sandy Lydon, a local historian and treasure, walked the entire coastline over a two-day period and wrote several articles for the Sentinel about his adventure. Other than needing an armed guard as an escort while he passed in front of the firing range at Ft. Ord, he completed the trek on his own and lived to write about it.

Five years ago in 2009, Sandy talked me into retracing his steps and bringing along 40 others who signed up for this adventure challenge through a Cabrillo Extension course. People were actually willing to pay money to hike 33 miles on sand over a two-day period! Fortunately we did this over two different weekends, which gave the participants a chance to nearly recover from the first day 15-mile ordeal. The agreement was that I would cover the first 200 million years of Earth history, geology and related trivia along the way, and Sandy would generously cover the last 200 years. I thought that sounded fair.

Day one would take us from our starting point at New Brighton State Beach to Moss Landing, and day two would start on the opposite side of the Moss Landing harbor and, with any luck, everyone would survive and we would get to Monterey before nightfall. Other than having to cross the nearly waist deep, cold water of the Pajaro River late in the first day, we arrived intact but tired at Moss Landing. Although I have to say, everyone was pretty beat.

The second day was a bit rougher going, in part because there was only one rest stop with a bathroom, at Marina State Beach. The bigger challenge, however, was the steep scalloped beach along Ft. Ord, which left everyone leaning over towards the ocean like they were at the Mystery Spot. By 6:00 that evening we had only reached the Monterey Beach Hotel, 3 miles short of our destination, and we were all dragging our feet. We decided to blow the whistle. Most of the group returned another weekend to finish the walk and earn our merit badge and certificate. We all discovered that a 15-mile hike on beach sand is a lot more challenging than 15 miles on a hard trail.

Adventure has been defined as gross discomfort seen in retrospect, and his trip fell squarely into that category. But it also provided a unique look at this bay we all call home from a perspective that most of never have the chance to experience. Walking the beach from Capitola to Monterey not only takes a lot longer than driving, about 17 hours longer, but also allows you the time to see and hear about what makes the bay so special. Sea otters and snowy plovers, sand mining and shipwrecks, seawalls and submarine canyons, and a history that includes gold mining, whaling, military bases and magnesium extraction.

Sandy and I had such a great recollection of our first trip that we decided to repeat it again in 2010. Forty brave souls signed on for the trek. Learning from our first experience, and to avoid complete mutiny, we broke the trip into three days: New Brighton to the Pajaro River, the Pajaro to the Salinas River, and from there to Monterey. A determined group hiked the entire shoreline and arrived tired but invigorated and inspired by what they had seen and accomplished.

Article 13

Walking Around the Bay

Sand City, where concrete dumped over the sandy bluff makes passage difficult. (© 2008 Gary Griggs)

In the last column I wrote about our trek around the bay, and while walking 33 miles in the soft sand over 3 days was a work out, it was also a rewarding educational experience for the whole group. After reading about our local marathon swimmer's attempt to swim across the bay, heading out in the middle of the night and being tortured by stinging jellyfish until she couldn't continue, I realized we took the easy route.

We had no pests to contend with, and we had a support vehicle that carried our lunches and cold drinks. Our dedicated drivers found access to the coast every few miles along the northern bay (the southern bay is a bit more remote), and could have rescued anyone needing assistance, but no one needed help. Everyone was motivated and had gotten some hiking in before hand. They also wanted to claim their certificate of completion and be able to show friends that they had completed the hike completely around the bay.

In an era when first-time adventure accomplishments are harder and harder to come up with, like 13-year olds climbing Mount Everest and attempting to sail around the world alone, our circumambulation of Monterey Bay may seem like a modest effort. But it's here in our backyard, it's doable without endangering your life, and when you arrive on the other side, you feel quite accomplished-tired but accomplished. You also have a new perspective on the bay's coastline and what's out there that you never see driving by on Highway One at 65 miles an hour.

There are actually only four places that present potential problems for a complete beach walk around the bay. You don't really have a choice at the entrance to Moss Landing harbor, unless you want to take up rough water swimming and dodge boaters and sea lions. But it's a not a bad hike along the edge of the slough, where we were fortunate to see a large raft of resident sea otters. It's about a mile diversion from Moss Landing State Beach, along Jetty Road, and then along Highway One, over the bridge and then out to the beach again. The only hazards are the cars and trucks honking at you, literally a few feet away.

On the left side of Highway One is the imposing Dynergy Power Plant where natural gas is converted to lots of electricity to power much of the central coast and bay area. The plant was originally constructed in 1949 and over the years it has been modified, rebuilt and upgraded, as well as bought and sold. Next to the power plant is the former Kaiser refractories where magnesium was first extracted from seawater to support metal production for the World War II effort, and then later for the production of high temperature bricks for steel furnaces. The plant closed in 1999, but Calera ("green cement for a blue planet") now uses the facilities as a test site for producing cement from seawater, eliminating the production of carbon dioxide that accompanies cement production.

There are three other potential barriers to the complete bay walk on the beach; two of these are wet ones. Depending upon the month of the year, the river flow and tidal stage, you may have to swim or wade across the Pajaro and Salinas river mouths. You may also be lucky and be able to walk across on a sand bar. On our first weekend of Bay Walk 2, the Pajaro was too deep to wade so we ended our first day's hike here. A week later the Salinas River ended in a sand spit, and we just boldly walked across. The Salinas River is actually barred up much more frequently than its open, simply because the summer river flows today are usually so low. Two dams and lots of agricultural usage take their toll. Well, that and a continuing drought.

One other obstruction is in Sand City where local concrete trucks poured their end-of-the-day dregs over the edge of the bluff for decades to form a very resistant stretch of bluff that now protrudes into the surf zone like the prow of a ship. It's a lot easier, however, to hike over this mass of concrete than risk your life trying to get between it and a breaking wave. When you get over this barrier, you only have about three miles left to go.

Article 14

Walking Around the Bay-Again

Camp McQuaide, Santa Cruz County c. 1948.
(courtesy: Monterey Bay Academy)

On three Saturdays in June 2012, Sandy Lydon and I led our 4th annual, 33-mile hike around the shoreline of Monterey Bay. We discovered the hard way on our first trek around the bay in 2009 that attempting the 33 miles on the sand in two days was asking a bit much of the 40 hikers, as well as ourselves. Shoot, Sandy and I were at the upper end of the hiker's age range.

So in the 2nd, 3rd and 4th years we decided to break the hike up into three nearly equal segments, each about 10-11 miles, essentially determined by natural barriers. Setting off from New

Brighton State Beach on June 2nd, we managed to again convince 40 adventurous people to join us.

New Brighton Beach was known for years as China Beach, from the small Chinese fishing village that was located on the back beach in the late 1800s. In 1880 there were 29 men living in the camp, all born in China, ranging in age from 16 to 43, and who fished in the offshore cove with nets. The summer residents or tourists either walked down into Capitola village to buy fish or waited until the Chinese peddlers with their fresh fish would come by their doors.

The first day's bay hike is so full of stories of history, geology and coastal disasters that by about 11:00 we had told a lot of stories but only covered 2 of our 11 miles. Between the repeated wave attack and seawall reconstruction at Seacliff State Beach, the life and times of the World War I concrete ship (its not really cement), the filling of the swampy Rio Del Mar Flats for homes, the construction of what was reportedly the world's largest fresh water swimming pool along the channeled mouth of Aptos Creek, and the creek's frequent desire to flow downcoast towards the homes built on The Island, there was no shortage of stories.

A short distance downcoast, the 1983 El Niño storm damage to homes and seawalls along Beach Drive and Via Gaviota have long since been repaired. Newer, larger and more expensive seawalls have been built to provide protection against future storm waves and high tides. So far, so good, although sea level continues to slowly rise.

We finally reached Manresa State Beach for lunch and tales of iron and gold mining from the Aptos beach sands, climate change and sea level oscillations, before we started the casual 5-mile afternoon stroll to the Pajaro River.

This portion of the hike follows a little less developed section of shoreline, although during World War II, the former Camp McQuaide housed hundreds of US Army deserters and other military troublemakers up on the bluff south of Manresa. It was also home to the 250th Coast Artillery Regiment, the Signal Corps,

Cavalry Medics and a naval radar group. The Camp even had a small landing strip so airplanes could take off and practice dropping bombs on offshore targets. When the war ended, Camp McQuaide was determined to be surplus and decommissioned.

Although the 400-acre site was considered for a community college, and there was local support, the location and highway access weren't ideal and building refurbishing was going to be costly. So with no interest from any government entity, the Seventh-Day Adventists, who subsequently founded the Monterey Bay Academy, purchased the 400 acres of ocean view land for one dollar! That has to be the land deal of the century.

Day One for our bay walk ended at the mouth of the Pajaro River. Littoral drift from the north nearly always pushes the river mouth south, which added an extra half-mile or so to our walk.

We started at the south side of the Pajaro River on the morning of June 9, at Zmudowski State Beach, a difficult to pronounce park that was named after Watsonville school teacher, Mary Zmudowski, who donated the land to the state in the 1950s. We carefully avoided any nesting Snowy Plovers and started on the 9-mile hike to the Salinas River mouth. This section of shoreline is fairly serene and little used, with the stacks of Moss Landing always appearing to be closer than they are.

You can't walk across the entrance to Moss Landing Harbor, so we had to leave the beach, hike out Jetty Road, which collapsed due to liquefaction during the 1989 earthquake. We then carefully walked along about a half-mile of Highway 1. After the morning solitude of a quiet and isolated stretch of beach, it's a shock to your senses to find yourself along the narrow shoulder of Highway 1 with cars and semis passing at 60 miles an hour a few feet away.

The state's largest thermal electric power plant looms on the left, next to what used to be Kaiser Refractories. For decades, magnesium was extracted from seawater for the production of material for bombs during World War II, and then for later for magnesium oxide bricks for high temperature steel furnaces. The factory was closed a number of years ago when importing the material from China became more economical.

We returned to the shoreline after crossing the Sandholdt Bridge, also damaged during the Loma Prieta earthquake. The new Moss Landing Marine Laboratories now sits comfortably on a hill, well inland from the sand spit where it was formerly located, another victim of seismic shaking and liquefaction in 1989. And the final 16 miles is the hard part.

Article 15

On the Beach
Moss Landing to Monterey

Concrete previously dumped over the bluffs, Sand City.
(© 2012 Gary Griggs)

The 2nd half of our circumambulation of Monterey Bay becomes a bit more demanding than the first half, more like Survivor Ft. Ord than a stroll on the beach. Day two from the Pajaro River mouth, through Moss Landing to the Salinas River, is the shortest of the three days and an easy hike - about 10 miles on a reasonably hard and flat beach.

Moss Landing, at the precise midpoint of the bay, has a long and interesting history from Norwegian whalers, to salt ponds and magnesium extraction, to marine research and education, and finally whole enchiladas.

While the Salinas River historically flowed northward behind the dunes for nearly six miles from its present mouth, to discharge north of the present day entrance to Moss Landing Harbor, farmers decided to shorten and straighten the river's course in about 1909 and apparently cut a channel straight to the sea. The river meanders around a little at the mouth today but has more or less behaved for a century.

In the mid-1940s, the Moss Landing drainage system was altered again. The Army Corps of Engineers dredged a channel through the dunes directly opposite the mouth of Elkhorn Slough and created a permanent entrance channel protected by jetties. While this created a more stable entrance channel and shortened the route to the sea for slough drainage, it also steepened the gradient and led to the beginning of down cutting and erosion in the slough.

The relative ease of Day 2 quickly became a distant memory as we left the Salinas National Wildlife Refuge. The beach gets progressively coarser, steeper and softer proceeding southward. For the next 8 miles or so, you can either walk along the steep, narrow and wet part of the beach, dodging the incoming waves, or walk along the dry, flatter and softer upper part of the beach (the berm), although either begins to take its toll on your body after several miles. The beaches along the southern bay shoreline are quite coarse, and as a result are steep and poorly compacted. Walking just takes a lot more energy because you sink deeper into the sand.

CEMEX is still sucking beach sand out of a pond on the back beach and dragging sand off the beach with huge buckets to keep their Marina sand mining business going. Removing the equivalent of 20,000 dump truck loads a year has to have an impact on the shoreline, and erosion of the shoreline southward from Marina to Monterey appears to be the direct effect of the continuing beach sand removal. For some reason, no permitting agency seems capable or willing to take the steps necessary to terminate this ongoing beach sand mining operation.

Another former sand mining site in Marina has now been converted to a resort in the dunes, right next door to the sewage treatment plant. Measurements from historic aerial photographs at this location taken between 1976 and 2004 indicate that the sandy bluff has been retreating at about 5 feet/year on average. There are going to be some challenges here in a few years.

Dune erosion a few miles to the south at Ft. Ord has continued at over six feet per year for decades. While Stilwell Hall was a special place for many World War II soldiers where they enjoyed their last evenings before heading overseas, bluff retreat led first to its abandonment and finally to its demolition in 2004. The large volume of broken concrete and rock that had been dumped over the bluff over the years in an attempt to slow erosion has all been removed and the beach has returned to a natural state.

Scattered along the beach fronting Ft. Ord are also the remains of 4 large concrete sewage outfall pipes, which originally discharged at the edge of the dunes, but the remaining support pilings stand today as sentinels about 200 feet from the bluff as erosion continues.

The coastline changes dramatically when you reach Sand City. For about half a mile of shoreline, the entire bluff consists of either broken concrete slabs that were dumped over the bluff edge, or what initially appears to be gray lava flows oozing down the slope, but which is actually the remains of decades of dumping the left over concrete from Redi-mix trucks. This has formed a very hard crust over the sandy bluff that resists wave attack. At one point, the resistant concrete forms a promontory that juts out into the surf, and is the only place (besides the entrance to Moss Landing Harbor) where you cannot continue a complete bay walk along the beach.

At this point we reached civilization. After miles of quiet and unpopulated sandy beach, where we encountered only a few dead seals and birds, and a plastic detergent bottle from Japan, people with towels and umbrellas all of a sudden begin to appear.

But we still have about three miles to go. Erosion threatens shoreline development as we continue down coast, evidenced by the recently rebuilt concrete seawall protecting the Monterey Beach Hotel, which becomes a peninsula in the winter months. And then there is the new overhanging concrete wall fronting the Ocean Harbor House condominiums, which was supposed to look like an eroded sand dune. We didn't think so. During the winter months, beach walkers cannot pass safely along the beach in front of this seawall at high tides.

But three days and 35 or so miles later, the group finds renewed energy and a feeling of accomplishment as we approach Fisherman's Wharf in Monterey, happy to be alive and still on our feet.

Article 16

Walking on Rocks

Colorful pebble beach, Kas, Turkey.
(© 2012 D Shrestha Ross)

In early October 2012, Sandy Lydon and I led an intrepid group of 40 other hikers were on Day 2 of our 5th Monterey Bay Walk. On day 2, we departed early from Zmudowski State Beach, just south of the Pajaro River mouth, for Marina State Beach, a 9-10 mile easy walk.

It's a wilder stretch of shoreline than Day 1, which took us a long 12 miles from New Brighton State Beach, past Seacliff Beach, Rio del Mar, Manresa, Sand Dollar, Sunset and Palm beaches and finally across the Pajaro River, dammed by a sand bar, and down to Zmudowski. Lots of back beach home development, but also state beaches, sun bathers, walkers, joggers, fishermen, and a few surfers.

But, the thing I think we all take for granted that makes this long walk possible, the walk to remember for some, is that we have a 31-mile long ribbon of sand extending along the entire bay shoreline. While the beach from New Brighton to Monterey ebbs and flows with the seasons, eroding back in winter and accreting in the summer, there is always sand to walk on. In fact, there are only two places where it is usually impossible to walk along the dry sand, sometimes four.

Nobody who has taken the walk yet has been able to walk across the entrance to Moss Landing Harbor, so a detour across the Highway One bridge is necessary. At the southern end of the bay as a result of concrete and debris dumped over the bluff at Sand City for years, there is now a pile of concrete extending out into the water. It's a difficult and dangerous place to walk around, so we climb over it.

Depending upon the time of year and river flow, you may be able to walk across the sand bars at the mouths of both the Salinas and Pajaro rivers, but then again you may not. In 2012 our group was able to walk across both as river flows were very low in October, as they usually are, and sand bars dammed the mouths.

A beach 31 miles long requires a lot of sand from somewhere. In round numbers, the shoreline from New Brighton to the Monterey breakwater contains about 20 million cubic yards of sand; that's about two million dump truck loads. And that's not counting all of the sand that has been blown up into the dunes along the central and southern part of the bay, stretching from Pajaro Dunes to Del Monte Beach, or that that underlies all of the Fort Ord and Sand City areas.

So where did it all come from? Mostly its come from the rivers and creeks between Half Moon Bay and the Salinas River. Their watersheds are underlain by granite, sandstone, siltstone and shale and a variety of other rocks, but if you look at the beds of most of the streams as they head into the ocean, they are transporting sand- nice clean beach sand.

Most of it is carried by the high winter river flows, and in fact, one winter with a lot of rain can bring more sand to the beaches than five or ten years of low flow - except where we have built dams.

We spent three weeks in Turkey in September 2012, before starting Bay Walk 5. An interesting place to travel, especially right now. We did spend time along the Aegean coast and generally, the water is clear and clean, but there are very few, nice soft sandy beaches. We didn't find a Monterey Bay shoreline.

Most of the beaches are gravel or cobbles, smooth and rounded, but very coarse. Not the sort of beach you would jog or play volleyball on, but beautiful beaches. The people don't seem to mind, and once you have hobbled across the beach and into the water, you scarcely notice.

You might ask why? But even if you don't I'm going to explain why. Most of southern Turkey is pretty arid, not a lot of rainfall, and not many large rivers carrying sand to the shoreline.

The coastline of southern Turkey also consists dominantly of marble and limestone, great for building those ancient Roman cities, the ruins of which are scattered across the landscape, but not so great for producing sand.

Marble and limestone consist of calcium carbonate, made of ancient plankton with calcareous skeletons, or coral in some cases. But when these rocks are eroded, by weathering or from wave attack of the cliffs, they break down into small pieces of limestone and marble, which eventually dissolve. They just don't stick around as sand size grains. So the water is clear but the beaches take some tough feet.

Article 17

Wilder's Fallen Arch

The natural arch, Wilder Ranch State Park, Santa Cruz County, c. 1900. (courtesy: Special Collections, University Library, UC Santa Cruz)

The coastal trail through Wilder Ranch is a great hike or bike ride, and it's virtually in our back yard. No matter how crowded Main Beach, Cowell or Natural Bridges may be, you can find solitude out along the cliffs a few miles away. You might see harbor seals hauled out on the rocks, and also a lot of natural history.

A century ago a large natural arch spanned the cove that was the Wilder Ranch dump, and which supplied the beach glass on nearby Fern Grotto Beach I wrote about earlier. This arch was photographed in about 1900 with not one, but two horses and buggies, perched on top of it. This wasn't a delicate arch, but a pretty meaty natural bridge that the buggy drivers clearly had some confidence in.

Based on the dates of historic aerial photographs, this grand arch collapsed somewhere between 1928 and 1943. There was also another lower and smaller arch just south of Fern Grotto Beach that survived until around 2000 or so.

Why are there arches here anyway? If you have looked at any of the old photographs of the coast between Lighthouse Point and

Collapsed arch, Wilder Ranch State Park, Santa Cruz County.
(© 2005 Gary Griggs)

Davenport, or explored many of these beaches, you will have seen a lot of natural bridges, arches, and caves. Natural Bridges State Beach is probably the best known and most visited, but there are and have been lots of others. Although there is only one bridge left at Natural Bridges today, the name may stick for a long time to come.

The book we did several years ago, The Santa Cruz Coast-Then and Now, includes a lot of these old photographs. And a lot of bridges have come and gone over the years. What is needed to create a natural bridge or arch is rock that is strong enough to stand vertically and form cliffs, but which also has internal weaknesses that the waves can selectively erode.

In many cases this may simply be a softer or weaker layer near the base of the cliff, where the waves can focus their energy. And you see a lot of undercutting at the base of the cliffs along the Wilder coastal trail, which is evidence of a weak layer or softer rock. There are also other sorts of weaknesses in the Santa Cruz Mudstone that makes up the cliffs - joints or fractures are places where the rock is more easily dislodged or removed by wave impact.

There are sandstone intrusions within the mudstone making up the cliffs (which I've written about in earlier articles), and these layers are typically easier to erode and may provide the focus for wave energy to begin to undercut the cliff. Given enough time, and the ocean has plenty of that to spare, over the course of tens or hundreds of year, caves can be deepened or expanded along a weak zone until some portion of the overlying rock collapses. Sometimes an arch remains, and some times it doesn't. The arch may last for decades, as those did at Wilder and Natural Bridges, or only a few years, as was the case of a large arch that formed at the south end of Four-Mile Beach in the 1980s.

A century ago there were arches on either side of Lighthouse Point, also one at Bird Rock, another at the end of Woodrow (the Vue de L'eau) and three at Natural Bridges. Today only one of these remains but there is newer arch at the west end of It's Beach that you can walk through at low tide.

Article 18

West Cliff-Stepping Back in Time

The Purisima Formation, sandwiched between terrace deposits
and Santa Cruz Mudstone, at Swift Street & West Cliff Drive, Santa Cruz.
(© 2014 Gary Griggs)

West Cliff to many, is pelicans and cormorants, seals and sea otters, and the occasional whale or dolphin if you're lucky. While I do notice the wildlife, I tend to focus on the rocks. And there are a lot of rocks with some interesting stories and a lot of history along West Cliff. The difference is that are always there, they don't move around, and they're really old.

Geologists think in terms of millions of years, although for most people, anything beyond a few thousand years is ancient history. When my field trip partner and occasional traveling companion,

Sandy Lydon, and I are leading hikes or speaking together, we usually divide things up by time and try not to tread too hard on each other's terrain.

I usually take the last 200 million years of Earth history and Sandy takes the last two hundred years of human history. Seems like a reasonable division of labor. Although he usually has a lot more to talk about than I do, perhaps because he has old newspapers to read and I just have the rocks to look at. But, the advantage is that I can make up stuff; although after many field treks together, I think he does that too.

So while I insert a little history into some of my columns, I often check with Sandy first, and he is quite generous in offering me a 2 or 3 paragraph correction of my erroneous information. Sticking with the rocks is a little safer for me.

In the larger geologic scheme of things, the rocks exposed along West Cliff are pretty young, adolescent really. There are just two main rock units out along the cliffs, with some very young beach sand and dirt scattered over the surface.

Scientists like to break down their subjects into workable units, no matter what the science. Biologists use species, genius, families, orders and a few others, while geologists use formations. These are sequences or large masses of rock that are distinct and recognizable over large areas, and represent some set of conditions or an environment that persisted for some lengthy period of time.

Along West Cliff we have two geologic formations that happen to meet each other along the sea cliffs between Swift Street and Almar Avenue. The cliffs extending from Swift Street north all the way to Waddell Bluffs, all consist of a rock known as the Santa Cruz Mudstone. These rocks are believed to have formed on an ancient seafloor about 7-9 million years ago.

The mudstone isn't a terribly exciting stack of rocks as rocks go. It consists mostly of hardened clay or mud, has very few visible fossils, but is nearly 9,000 feet thick. The mud was originally deposited on an ancient sea floor some distance offshore when this entire area was below sea level. The mudstone does contain a lot of silica from planktonic organisms like diatoms, which made siliceous skeletons in that ancient ocean. The mudstone also has some interesting weathering patterns and other somewhat unique structures.

Years of burial and compaction squeezed out the water and slowly converted the mud to mudstone. The tectonic activity or mountain building along the edge of California in the not too distant past slowly pushed up the Santa Cruz Mountains and lifted the mudstone above sea level where we find it today. Because of the high silica content, the Santa Cruz Mudstone is generally hard and somewhat resistant to wave erosion. It stands in vertical cliffs 30 to 50 feet high or even higher between Scott(s) and Waddell creeks.

The arches at Natural Bridges State Park were carved into the mudstone, and as you proceed along West Cliff towards the lighthouse, you are walking on this formation. However, something special happens at the end of Swift Street, which the casual observer would most likely miss, but that hopefully you won't.

Resting on top of the Santa Cruz Mudstone and exposed on the bedrock platform just below the sidewalk at the end of Swift Street, is the next youngest sequence of rocks, the Purisima Formation, just a few inches thick here. This roughly 7 million year old formation is a far more interesting sequence of sedimentary rocks. And perhaps without knowing it, many locals have a chunk of the Purisima, filled with mollusk shells or perhaps a piece of bone, sitting on their bookshelf, windowsill or front porch.

Article 19

The Purisima - From Clams to Cetaceans

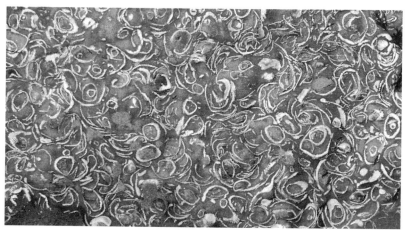

Fossils in the Purisima Formation, Depot Hill, Capitola.
(© 2007 Gary Griggs)

The Santa Cruz Mudstone, which extends north along the coast from West Cliff to Davenport and beyond, is pretty boring and monotonous as far as rocks go. There's just not a lot to see in the mudstone, with the exception of some of the world's largest sandstone intrusions at Yellow Bank Beach. The Purisima Formation, on the other hand, particularly the section from Pleasure Point east and continuing through Capitola and into the bay, has about as much as you could want in a local stack of rocks.

For geologists, it's mostly all about how, where, and when did these rocks form. We spend our lives asking and then trying to answer what seem like relatively simple questions. And as many times as I've looked at the fossil beds and whale bones in the Purisima, the answers still aren't completely clear to me.

Part II: Coastal Geology & Field Trips

Geologists have worked from the principle of uniformitarianism, or "the present is the key to the past", for perhaps 200 years. I think it might be the opposite for human history, where the past hopefully provides some insight to the present.

So if we see an outcrop of rocks, perhaps in a road cut or a sea cliff, and we notice the grain size and layering of the sediment looks the same as we might see in a modern sand dune, we can reasonably conclude that these ancient layers exposed in the road cut were deposited by wind in the past. The sand dunes of the present inform us about the ancient preserved dunes of the past.

The Purisima Formation contains a lot of clues about its history. The layers of clam shells exposed in the 70-foot high cliffs of Depot Hill, and the scattered cetacean (whales in some cases) bones that are also commonly found in rocks that have fallen from the cliffs, provide some evidence. You may also recall several years ago when the skeleton of a small whale was found embedded in intertidal rocks during the construction of the seawall along East Cliff near Pleasure Point.

Take a trip out to the Seymour Center at Long Marine Lab, and notice the three large slabs of Purisima bedrock right in front of the world's largest blue whale skeleton. Each contains vertebrae and ribs of a 7 million year old whale, or perhaps several whales. These boulders were collected below the cliffs at Depot Hill and give us some sense of what this area was like in the not too distant (geologically speaking) past.

So playing detective, where would you look today to find an environment where huge numbers of clams lived and died, and also where whales might occasionally die and sink to the seafloor to be buried in the sediment? It was a marine environment, probably where water was not terribly deep, perhaps not that different than Monterey Bay today. We got clams and recently, lots of whales.

The other question to answer is why did so many clam shells end up in these particular layers, and why did the whales choose to die here? Well, good questions. Where might we find remains like these today? We do see lots of clamshells scattered along the beach at low tide along some Monterey Bay beaches, and also concentrations of shells along tidal channels in places like Elkhorn Slough or Francisco Bay. But the whales? I'm open to ideas on this one.

If we look at the exposures or outcrops of the Purisima Formation today, this gives us some idea of the area that was occupied by the former Purisima Sea 7 million or so years ago. These rocks extend along the coast from the end of Swift Street on the west side to Aptos Seascape to the southeast. They also extend 6 to 8 miles inland and underlie most of the watersheds of Soquel and Aptos Creeks.

It's the Purisima Formation, these fossil-rich sandstones and mudstones, which supply almost all of the water to the Soquel Creek Water District. Rainfall and stream flow, percolating down from the surface has been stored in the pore spaces between all of the sand grains of the rock over thousands of years. We have been pumping out that groundwater in recent years at faster rates than it is being recharged or refilled. This has become a common problem in California, and elsewhere, but obviously can't go on forever.

Fossil whale bones in the Purisima Formation, Seymour Marine Discovery Center, Santa Cruz.
(© 2007 Gary Griggs)

Article 20

The California Coastal Records Project

As Californians we have 1100 miles of coastline as our own ocean backyard. If we divided up the state's coast equally between our 38 million residents, we would each have less than two inches. Residents of Nevada and Arizona have a lot of sand but no coastline. Along the central coast you can look out the car window as you drive Highway 1 and see almost the entire coast from Morro Bay to San Francisco.

But there is much of our coast that is nearly impossible to see firsthand; large sections of Humboldt and Mendocino counties come to mind. Places like the Cape Mendocino and the Lost Coast are very difficult to get to due to the lack of roads.

There is a partial solution, however, thanks to two county residents who took it upon themselves to document the entire coast in photographs. Twelve years ago, they were listening to a conversation about the difficulty of monitoring change along the shoreline, such as unpermitted construction or other land use modifications.

Kenneth and Gabrielle Adelman happened to own a helicopter and decided in 2002 to fly and photograph the entire coast from the Oregon border to Mexico, realizing that their photographs would form a permanent record of what was there at that moment in time. While Gabrielle flew, Kenneth took thousands of high- resolution digital images of the entire coastline and then put these up on a publicly accessible website (The California Coastal Records Project).

While the site might have gone unnoticed for a while, it received attention very quickly. Barbra Streisand, who lived on a cliff top in Malibu and who liked to be thought of as an environmentalist, objected that one of the 12,000 photographs on the site, happened to include her house. She decided to file a lawsuit against Kenneth and Gabrielle claiming they were invading her privacy.

Streisand lost the case but the media coverage caught my attention and I clicked on the site to see what the big deal was. I immediately realized what a valuable resource this was, not only for someone studying the shoreline, but also for anyone with an interest in California's coast.

I also happened to have in my office nearly 14,000 35mm color slides taken of the entire coast by a coastal engineer working for a state agency in 1972 and 1979. I contacted the Adelmans, who expressed interest in adding these to the site. This initiated a wonderful collaboration and led to their assistance in scanning each of these slides, correcting each slide for the faded colors from 30 years of sitting in boxes, and posting them on the site.

They have now flown the California coast 7 additional times, in 2004, 2005, 2006, 2008, 2009, 2010 and 2013, and also obtained several other sets of older photographs, which they have scanned and added to the site. There are now nearly 110,000 high-resolution digital images of the coast of California, extending in some places back to 1928 that can be easily viewed by anyone, anywhere in the world, thanks to their dedication.

We decided to pull together some of the best of these photographs, write a description or some history and explanation about each of the locations selected, and put these together in a book. California Coast from the Air-Images of a Changing Landscape, was published in early 2014, and is now available at the Seymour Marine Discovery Center, other bookstores, and online.

PART III

COASTAL EROSION, PROTECTION & SHORELINE CHANGE

Article 1

California Coast-Worn at the Edges

Storm waves hitting Lighthouse Point, Santa Cruz, 2008.
(© Shmuel Thaler, Santa Cruz Sentinel)

At only 20 or 25 feet high, the waves from a mid-January 2010 storm didn't allow the organizers of the Maverick's Big Wave contest to blow the starting horn. They have decided to wait for larger waves. A month earlier and a few miles north, residents of an apartment complex in Pacifica were evacuated as storm waves tore away at the sandy bluffs beneath their homes. Although a contractor was hired to pile rocks at the base of the bluff, at least some of the residents have lost interest in moving back in. How long can a pile of rocks hold back the Pacific Ocean? While the surfers are hoping for larger waves, some ocean front residents have had enough winter already. They don't share the same excitement for 30 to 50 foot high waves.

Large waves arriving at high tide are the major force behind most coastal erosion and storm damage. Shoreline retreat may take several different forms, however. As described in an earlier column, the change from low-energy summer waves to high-energy winter waves leads to beach erosion every winter. Sand is scoured off the beach in December and January and stored offshore, only to return again the next spring and summer when winter storms have abated and calmer waves return. This is an expected and normal process we can all observe. The severity of winter storm waves varies from year to year, however, usually being greater during El Niño winters. As a result, the extent of beach retreat and recovery is different each year as well. All oceanographic signs in early 2010 pointed towards a moderate strength El Niño, but this was of little comfort to coastal residents.

The coastal erosion that concerns cliff top residents as well as coastal communities isn't the seasonal beach erosion, however, but the erosion and permanent retreat of the cliff or bluff. This is not recoverable, at least within our lifetimes or by natural processes.

The rate at which cliffs or bluffs have historically eroded along the California coast varies from a surprising ten feet per year at some unfortunate locations, to a few inches or less in others. The former is obviously a cause of concern, especially if it's your front yard or patio. Several different factors affect how rapidly a given cliff will retreat. The strength or resistance of the material making up the cliff or bluff to wave attack is usually the most important. Comparing West Cliff and East Cliff, the rock strength varies widely, depending on the type of rock, the hardness or degree of consolidation or cementation, and the presence of weaknesses such as fractures or joints.

The amount of wave energy reaching any particular area of coastline is also a key factor, and while the waves approaching the central coast on any given day come from the same storms, nearshore bottom conditions or bathymetry will increase or decrease wave heights at specific locations along the shoreline. Waves at

Steamer Lane are always going be larger than those breaking on Main Beach or at Capitola. The final factor, the regional rate of sea-level rise will need some additional explanation.

Article 2

Broken Bridges and Fallen Arches

Arch at Steamer Lane, Santa Cruz, c. 1888.
(courtesy: Special Collections, University Library, UC Santa Cruz)

The coastline of Santa Cruz has undergone some dramatic changes over the past century or so, and watching the waves batter the cliffs this past month it's easy to understand why. As coastal geologists we often take careful measurements from old aerial photographs to see how much retreat has taken place over time and then calculate cliff retreat rates. For the much of the Santa Cruz coastline the average annual erosion rates typically range from a few inches to about a foot per year. Cliff failure doesn't occur in simple six inch or one-foot increments every year, however. Instead we see arches collapse catastrophically, or as happened in Depot Hill two weeks ago, slabs four to six feet wide fail instantaneously, removing a few more fragments of the former Grand Avenue and encroaching a little closer to the cliff top homes.

Several years ago we published a book comparing old photographs of the Santa Cruz coast, many taken 75 to 100 years ago, with photographs we took from the same locations in 2006 ("The Santa Cruz Coast-Then and Now"). The arches of West Cliff Drive fascinated early visitors as they do today, but the same rock weaknesses that allowed the waves to erode those arches, have also led to their demise. Natural Bridges (previously known as Moore's Beach, Hall's Beach, and Swanton Beach) has probably been the most widely visited and photographed. Richard Hall, who came from Vermont to Santa Cruz in 1853, bought 300 acres on the Cliff Road, which included the beach and "three natural bridges". Historical photographs of Natural Bridges indicate that the outermost arch was intact in the fall of 1905 but collapsed shortly thereafter. The inner arch lasted for at least 125 years but failed during a storm on the night of January 10, 1980. Today only a single arch remains.

Interestingly, a half-mile to the west, directly in front of the Seymour Marine Discovery Center at Long Marine Laboratory, the mast of a coastal vessel (the La Feliz) leans up against the cliff. The La Feliz was washed onto the rocks here during a severe storm on October 1, 1924, and ninety years later, the mast and cliff are virtually unchanged.

Further east at the end of Woodrow Avenue (formerly Garfield Street), there was a triple arch for a while. This was a popular site for photographs in the late 1800's and early 1900's. Part of this arch ultimately collapsed, leaving the famous Vue de L'eau, a picturesque single arch memorialized on many old postcards. Bird Rock, a safe haven today for countless cormorants and pelicans, was connected to the coast by an arch that collapsed sometime in the 1920's.

The east side of Lighthouse Point was the site of another arch where photographs of ladies in long Victorian dresses and strange hats were often taken. A large storm in the winter of 1888 finally brought down the arch, but its base still stands in shallow water today, directly in front of the stairway where surfers heading to Steamer Lane enter the water. Lighthouse Point is now partially undermined by two separate caves, which in time will collapse, perhaps leaving new arches for the photographers of the future.

Natural Bridges State Beach, c. 1890.
(courtesy: Special Collections, University Library, UC Santa Cruz)

Article 3
Migrating Shorelines

Opal Cliffs from 41st Avenue to Capitola, 1928.
(photo: former Ocean Shore Railway, Gary Griggs Collection)

From the long perspective of geologic time, the coastline of California, like the coastline of any area on Earth, is a very temporary feature. Twenty thousand years ago, just a blink of an eye in the long-term scheme of things, the shoreline near Lighthouse Point was about ten miles offshore to the west. The Farallons were a 30-mile hike across a flat, sandy plain. In central and southern Monterey Bay, the sand that was being brought to the shoreline by the Salinas and Pajaro rivers was deposited five to ten miles seaward from the present beach along the exposed shelf. Onshore winds blew this sand into large dunes, which extend inland today from Sunset Beach to Monterey and underlie the old Fort Ord army base.

As the last Ice Age drew to a close, the weather gradually began to warm, and the glaciers that covered the upper mid-west, Canada, Alaska and northern Europe, began to retreat. As they melted, all of that ice water poured into the oceans, eventually raising sea level globally about 400 feet. The first humans slipped across the Bering Land Bridge into North America before sea level rose too high, and began their trek southward. Not long ago, cave divers in the Yucatan discovered the oldest human skull found to date in the Americas, that of a young girl, which was dated at about 13,000 years old.

If you were one of those early human arrivals that populated coastal California, you would have simply moved inland to keep your feet dry and your camp above the level of the slowly advancing ocean. During the period from about 18,000 to 8,000 years ago, sea level rose vertically 375 feet, or almost half an inch a year. In response, the shoreline advanced inland a little over five feet each year towards present day Santa Cruz. Off the Golden Gate where the seafloor slopes more gradually, the shoreline was advancing at over fifteen feet each year. This didn't matter much in those days, but at this rate, we would completely lose Main Beach in 15 to 20 years. The Pacific Ocean is 10,000 miles wide and a few miles one way or other at the edge doesn't really matter too much to the ocean.

Shoreline migration or coastal retreat wasn't a concern in the Monterey Bay area until about the last 100 years. Older civilizations elsewhere around the world did have to deal with the slow rise in sea level, however, and marine archeologists regularly come across the drowned harbors and the sunken remains of ancient settlements around the margins of the Mediterranean.

In Santa Cruz, however, the first homes were set well back from the shoreline. The oldest aerial photographs of the Santa Cruz coast were taken in 1928 and show that homes weren't built on the cliff edge in those days. There wasn't a house within a block of the cliff edge along Opal Cliffs in 1928. Ocean views either weren't in the same demand as they are today, or the earlier coastal residents had a greater respect for the temporary position of the coastline than the builders and homeowners of the last half of the 20th century. Sea level has gone up and down and coastlines have advanced and retreated for as long as we have had an Earth and an ocean. The consequences of a rising sea level are just a lot greater today than at any time in the geologic past.

Article 4

Hazards of Living on the Edge

Postcard depicting Lover's Lane, Capitola, c. 1906.
(courtesy: Frank Perry)

Many of you may have noticed that there are always many more For Sale signs along our coastline in late spring and summer than during the winter months. Ocean front homes just look a lot more appealing to potential buyers in June, July and August than in December and January. Some very creative words are often used to describe those shoreline homes in the real estate ads: "On the Sand" and "Steps to the Beach", being good examples. Generally speaking, being "on the sand" and only "steps to the beach" is probably not the best place to invest your life savings. There are lots of exciting stories and photographs of serious storm damage and even destruction for homes built on the sand in California.

There are also some enticing street names dreamed up by coastal developers in the past. Malibu has Sea Level Drive, which raises an interesting question. In the Del Monte Beach area of Monterey, you can buy a home on Spray Avenue, Surf Way, or Tide Avenue. There are also some older parcels and streets in that subdivision that are now completely underwater as the shoreline continues to retreat at a foot or two feet each year.

Grand Avenue in Capitola used to extend along almost the entire length of Depot Hill. In the early 1900's there was also a double row of trees and a sidewalk between the trees along the ocean side of the street, which was known at the time as Lover's Lane. Today the trees and most of Grand Avenue are gone, along with one house, six apartment buildings, and four parcels, all casualties of the progressive erosion of the coastline over the past century.

Measurements from old maps and aerial photographs chart a retreat rate averaging about a foot per year. Proposals to protect the 70-foot high cliffs of Depot Hill from further erosion have come and gone over the past 40 years but none has ever come to fruition. This is a tough place to hold the line. It takes a majority of property owners to agree on an approach, and then a lot of money. More importantly, any protection project requires approvals from the city of Capitola, and more challenging, the Coastal Commission.

Opal Cliffs is another difficult area for local homeowners whose backyards continue to get smaller year after year. From 41st Avenue eastward nearly to the overlook above the Capitola pier, dozens of residents living along Opal Cliffs Drive no doubt look forward to the calmer summer weather and a respite from winter wave attack. The views are great but there is the constant threat of the next slab of Purisima bedrock giving way and falling to the shoreline below.

West Cliff Drive is a different story. Much of the underlying bedrock is more resistant Santa Cruz Mudstone, but more importantly, with a single exception, the homes are all on the inland side of the street. Although much of this coastline has now been armored with riprap or large rocks, when cliff failure does take place, it's the pathway and West Cliff Drive that are threatened, rather than someone's backyard, patio, or home.

ARTICLE 5

Dealing with a Retreating Coastline

Cliff failure and loss of homes at Gleason Beach, Sonoma County.
(© 2005 Kenneth & Gabrielle Adelman,
California Coastal Records Project, www.californiacoastline.org)

The beach at Santa Cruz was a 10-mile hike to the west 18,000 years ago at the end of the last Ice Age. As glaciers retreated and ice sheets melted over the following centuries, sea level rose about 400 feet and gradually flooded the edge of California, moving the beach inland. The coast retreated about 300 feet per century during that era of warming and melting. Throughout this 18,000-year period, the sea cliffs marching back three feet every year didn't matter much. Although the Ohlones used the coast, harvesting fish and shellfish and hanging out on the beach, they didn't have permanent dwellings so the precise location of the beach and sea cliff didn't really matter.

Things are different today. The shoreline and sea cliff from San Diego to Santa Barbara is almost completely urbanized. In Santa Cruz County, homes, businesses, roads, parks and parking lots cover the coastline from Manresa to Natural Bridges. It's probably fair to say that California's most valuable real estate is right on the edge. But that edge is never in the same place for more than a few decades. The shoreline moves back and forth over millennia as sea level rises and falls in response to climate change.

Around the Mediterranean Sea, civilizations have dealt with this phenomenon for thousands of years. In California, however, our coastal development and construction history is much more recent. Photographs of coastal bluffs in Santa Cruz County from the late 1800's reveal that people didn't build right at the edge. But today, it's pretty much continuous development of one sort or another. The closer to that ocean view, the more valuable the house and land.

So how have we dealt with the erosion and retreat over the past 50-100 years and what are we going to do in the future? This is a messy and emotional issue involving expensive and difficult questions, and its not going to get any easier in the future.

Historically there have been three basic options for dealing with coastal retreat: 1. retreat or relocation; 2. armor or protection; or 3. beach nourishment. None of these are simple solutions so it will take a few columns to tell the whole story.

Nobody with an oceanfront location is excited about moving back from the edge, but it has happened and will likely happen more often in the future. On Depot Hill in Capitola, six cliff top apartments were taken down after the Loma Prieta earthquake when concrete caissons were undercut, and foundations cracked and partially failed. Twenty years earlier, a house next to the apartments was picked up and moved back several blocks and put on a new foundation.

In other cases, failure to relocate houses led to cliff collapse and homes ending up on the beach below, which is what happened along the Esplanade in Pacifica in 1998. Sea-level rise and a more severe storm climate claimed 28 villages along the southeast coast of England during the Middle Ages that are chronicled in a book written over a century ago, The Lost Towns of the Yorkshire Coast. Perhaps the most expensive relocation to date was the Cape Hatteras lighthouse. The National Park Service moved this 130-year old, 21-story tall structure, which weighs 4830 tons, over a half a mile inland on giant rollers in 1999 at a cost of $12 million. Coastal erosion and retreat are not really new issues, we just have a lot more at stake today.

Article 6

Perils of Paradise - Living at the Edge

Eroding coastal cliffs at Depot Hill, Capitola.
(© 2006 D Shrestha Ross)

Human civilization developed over the past 8,000 years or so during a time when sea level was nearly stable, or rising very slowly. There are some historians and archaeologists who believe that this nearly stable sea level, in contrast to the previous 15,000 years of rapid rise, may have played an important role in allowing the early agricultural civilizations of the coastal plains and deltas of the Eastern Mediterranean and Middle East, as well as China, to develop and flourish. There was some stability to the shoreline. It wasn't until several thousand years later, however, that ancient cities and harbors began to emerge along the eastern Mediterranean coastline.

There are lots of reasons why people then, as well as today, choose to live along the coastline: a more moderate climate, fertile flood plain soils for agriculture, and easy access to the sea for boats and trade among others.

As sea-level rise slowed from 3 to 4 feet a century to a less than a foot, it made a big difference along a flat, low-lying coastline. The coastal plain of the Carolinas, for example, typically slopes at about 1:500, which means that the land only rises a foot in elevation for every 500 feet you move inland. With a rise in sea level of 3 or 4 feet/century, the shoreline will move 1,500 or 2,000 feet farther inland.

While average global sea level only rose about 8 inches over the past century, along the sandy shoreline of North Carolina the rise varied from 10-18 inches, which took its toll. The Cape Hatteras lighthouse was originally built 1,500 feet from the shoreline in 1870, in what was considered a safe spot at the time. A century later, the waves were breaking near the base of the lighthouse. Almost 1,400 feet of shoreline retreat had taken place in 100 years, or 14 feet per year, on average. For a coastal property owner, 14 feet a year of property loss can be troublesome.

Relocation of buildings hasn't been the most popular historic approach for dealing with a retreating coastline, however, and has at least until recently been seen more as a last ditch effort. Armor or protecting the coast with some hard structure, a seawall or rock revetment, has been the typical response for most of the past century. What is amazing is that some of California's oldest seawalls are still standing. The concrete seawall protecting the old sewage pumping station just below the Grand Avenue Apartments on Depot Hill was built over 80 years ago. The O'Shaughnessy seawall along Ocean Beach in San Francisco was originally constructed in 1929 and has protected this stretch of the Great Highway ever since. O'Shaughnessy was a smart and methodical engineer who thought of every possible way in which a seawall might fail and then designed the structure to survive all of those possibilities.

We have seawalls in California that have survived 80 years and others that didn't make it through the first 80 days. Throughout the last century, our primary concern along the California coast was how to design and build structures that would protect coastal development and infrastructure from severe wave attack. And we have tried just about everything to hold off the Pacific Ocean with varying degrees of success. But we need to keep in mind that the Pacific Ocean is 10,000 miles across and it doesn't care too much about a few feet on either side.

Article 7

Seacliffs and Seawalls

Cliffs Hotel, Pismo Beach..
(© 2005 Kenneth & Gabrielle Adelman,
California Coastal Records Project, www.californiacoastline.org)

Seawalls are one of the most polarizing elements of our human interaction with the coastline. We seem to have a love-hate relationship with armor depending upon our perspective on the coast. A former chair of the California Coastal Commission stated many years ago that in the Commission's early years they often received requests of two sorts. One group who claimed that their oceanfront property was as stable as the Rock of Gibralter, that it hadn't eroded an inch in historic time, and therefore, building their cliff top dream home was perfectly safe. The second group believed that their oceanfront property was the most rapidly eroding in California, that their house was threatened, and that if didn't get a permit to build a seawall that they were going to lose their home in the next storm. The Commission chair's final observation was that quite frequently these two property owners lived next door to each other.

In order to assess the threat of coastal cliff retreat to an existing structure, which is what the Coastal Commission wants to know before making any decision on armoring, we first need to determine how fast a cliff is eroding. And perhaps to state the obvious, wherever we see vertical or near vertical sea cliffs, such as along West Cliff, Opal Cliffs and Depot Hill, often with large blocks of rock on the beach below, you can be pretty certain that these cliffs are actively eroding.

The question, however, isn't simply are they eroding, but how fast and does this pose an immediate risk to a structure? While this may seem straightforward to determine, it often isn't. We need an accurate record of how the position of the cliff edge has changed over time, ideally over 50 to 75 years or more. Because cliff failure tends to be an episodic process, where large blocks fail instantaneously, the longer the time period, the more representative our information will be. There may be no large failures or retreat for 10 or 15 years, but that short period of observations or data may not be typical of long-term conditions. And if you are going to invest your life savings in an oceanfront house, the expected lifespan of the property is worth looking into. No one wants to see their home dangling over the cliff edge, but there are plenty of people who have seen this happen. There are lots of structures and neighborhoods along California's coast that have literally disappeared into the sea.

The most common sources of information geologists use are historical vertical aerial photographs or survey maps. We need to find the oldest available maps or photographs, determine their scales, and then with the right tools and lots of experience, determine how much retreat has taken place over time. Different consultants looking at the same property may come up with different answers, which confuses everyone. Why different erosion rates for the same cliff? The scale and quality of the photographs or maps, the nature of the cliff edge and whether or not it is covered with vegetation or clearly exposed, the skill of the person conducting the investigation, and perhaps the reputation or ethics of the consultant can also play a role.

The proposal to build the Cliffs Hotel in Pismo Beach is a good example of this problem. When proposed in 1983 the applicant's geological analysis indicated an erosion rate of 3 inches per year for the 75-foot high cliff. A 100-foot setback from the cliff edge was required, which was more than adequate to provide 100 years of stability. The hotel was approved and built. Thirteen years later the owners returned and asked for approval to build a rock revetment for protection. A second geologic report was submitted based on 18 months of new data, which now indicated that the erosion rate was not 3 inches per year but 4 feet per year! At this rate the hotel would be threatened in less than 20 years. This is one example of a problem that both cliff top homeowners and the Coastal Commission must deal with on almost a daily basis. As my good friend and local historian Sandy Lydon would say: "And you call this a science"?

ARTICLE 8

Armoring the Coast

Winter wave damage to an ocean front home at Aptos, Seascape, 1983.
(© 1983 Gary Griggs)

During the first three months of 1983, the entire coast of California was hammered by the most severe El Niño storms in perhaps half a century. Twelve large storms hit the coastline and all but three of these arrived at times of high tides. Sea levels were a foot to nearly two feet higher than predicted along the state's coastline due to a combination of warmer water, storm surge, and an El Niño driven bulge in the sea surface that moved north from the equatorial region. The combination of an elevated sea level, high tides and large storm waves inflicted over $230 million in losses (in 2013 dollars) as houses, mobile home parks, commercial buildings, parks, harbors and public infrastructure were damaged or destroyed.

In low-lying downtown Capitola, waves washed into the beachfront Venetian Court condominiums and the restaurants along the Esplanade as debris and seawater were carried a block inland. Along the northern bay shoreline, homes were damaged along Rio Del Mar's Beach Drive. Storm waves overtopped the rock revetment protecting the homes along Via Gaviota in Seascape, broke through sliding glass doors and washed into oceanfront living rooms. Much of the newly completed timber seawall at Seacliff State Beach was destroyed, for the eighth time in 60 years.

Although many miles of seawalls and riprap already existed along California's coastline, the 1983 winter storm damage generated a flurry of permit requests from Crescent City to San Diego to rebuild old seawalls that were damaged or destroyed during the winter storms or to construct new ones. While city and county planning departments, as well as the Coastal Commission, subsequently issued many permits, concerns also began to be voiced about the effects of additional seawalls and riprap on the shoreline. The decision to allow armor or not has become an increasingly contentious issue in hearing rooms from one end of the state to the other. Ten percent, or 110 miles of the state's 1100-mile coastline has now been armored. The highest concentration of seawalls and riprap are in southern California, however, where one-third of the entire 233-mile coastline of Ventura, Los Angeles, Orange and San Diego counties has now been armored.

At the Coastal Commission level, the guiding language in the original Coastal Act of 1976 was somewhat ambiguous as it related to seawalls. One section stated that new development was not to be dependent on the construction of shoreline protection devices such as seawalls or riprap. Language elsewhere in the Act declared that seawalls and revetments shall be approved if an existing development is threatened by erosion. Attorneys, property owners and Coastal Commission staff have been debating ever since as to what constitutes "an existing development": existing at the time the Act was approved, or existing when the permit was requested?

Today permits for new seawalls have become the exception rather than the rule. Every proposal for a new seawall undergoes exhaustive scrutiny. Larger structures, such as the East Cliff Drive/Pleasure Point bluff stabilization project completed several years ago, typically go through an extensive Environmental Impact Report process. In that project's 280 page final EIR, which stretched out over 8 years, visual, biotic, hydrologic, geologic, recreational, traffic, cultural, and utility issues and impacts were all exhaustively analyzed. After considerable public input and review, as well as several hearings, the Coastal Commission approved the project based on the public benefits and the mitigation of significant impacts.

Article 9

Return to Pleasure Point

Pleasure Point prior to armoring.
(© 2012 Gary Griggs)

Steamer Lane and Pleasure Point both enjoy almost sacred status in Santa Cruz. Like Malibu and Rincon, they are known widely in the California surfing community and beyond as having nearly ideal waves, and a lot of surfers. It is not surprising then that any proposal or plan to alter either area would be met with some resistance and opposition. The cliffs at Steamer Lane have experienced both erosion and protection for decades. There used to be two Seal Rocks, but one has now disappeared beneath the waves. The rock pedestal on the beach in front of the stairs where some surfers get down to the water is the base of a much-photographed arch that collapsed during the winter of 1888. The stairway traverses a rock revetment installed in the mid-1960s, which has held up well to wave attack for nearly 50 years.

During the 1970s, a concrete plug was placed in the cave in front of the lighthouse in an effort to halt undermining of the point by wave attack and protect public access, but this has gradually succumbed to the impact of winter storms. Perhaps 15 years ago, the City Parks and Recreation Department briefly contemplated building a rock revetment which would have filled the cave and extended well out into the area where surfers take off. Wisely, this project was quickly abandoned. Eventually, this cave and the much deeper cave on the west side of the point will collapse and the point will change again. It's all part of the gradual retreat of the cliffs we see along the Santa Cruz coastline. Hopefully, nobody will be standing on top of it when it does.

Pleasure Point has faced similar challenges. The bedrock in both areas is the Purisima, a relatively young (geologically speaking) rock formation consisting of interbedded sandstone, siltstone and mudstone. While there are zones or layers of this formation that are harder and more resistant to wave attack, Lighthouse Point, San Lorenzo Point, Black Point, and Pleasure Point, for example, overall the Purisima is a relatively weak rock. By the way, are there any thoughts on why it is called Pleasure Point? The name took hold in the 1930s, gradually replacing the formal name, Soquel Point. One unconfirmed story reported in Donald Clark's book, Santa Cruz County Place Names, is that it got its name from a house of pleasure in the area.

While we are asking questions, how about Black Point? Early sailors described it as much darker than the surrounding cliffs and initially gave it a hybrid name of Prieta Point (prieta being Spanish for "blackish" or dark).

The stretch of cliffs from Pleasure Point to New Brighton Beach has almost no natural beach to protect the bluffs from wave attack, the exception being at Capitola, which owes much of its beach to a groin constructed in 1969. So the combination of weak bedrock and no beach to buffer the cliffs from wave attack has led to average long-term erosion rates along this stretch of East Cliff Drive of nearly a foot/year. Over the years as failure continued, there were scattered efforts to protect the cliffs and the roadway down coast of Pleasure Point. Loose rock and broken concrete were dumped on the beach from time to time, although there doesn't seem to be any record of who did this and when. Several concrete crib walls (think Lincoln logs made of concrete) were also built to support the roadway from collapsing as bluff erosion continued.

As the cribbing gradually failed, East Cliff Drive was threatened, and ultimately reduced to one lane. This also narrowed the space for bikes and pedestrians. In addition, water and sewer lines beneath the roadway were also getting closer and closer to being undermined. These concerns led the Santa Cruz County Redevelopment Agency about 14 years ago to look at a project that could stabilize the road, improve pedestrian and bike access, add a few parking places, clean the rock and concrete debris off the beach, and provide improved access for surfers and others wanting to get down to the beach safely.

Article 10

Protecting Pleasure Point

Pleasure Point Seawall.
(© 2012 Gary Griggs)

The ongoing retreat of the bluffs at Pleasure Point and the closure of one lane of traffic, threats to the water and sewer line beneath the roadway, increasingly unsafe conditions for pedestrians and bicycles, as well as the desire to improve public access, all led the County Redevelopment Agency to look at options for stabilizing this stretch of coastline about 10 years ago.

The project planning, engineering, environmental assessment and review, public input and meetings, and Coastal Commission review and hearings stretched out over five or six years. In this location, there were essentially only two options: try to stabilize the bluff to slow or halt the erosion, or do nothing and let the retreat continue.

The Environmental Impact Report or EIR had to look at several different protection options as well as the "no project" approach. If nothing were done in this area, where erosion proceeds at about a foot per year on average, it would only be a matter of time before more bluff collapsed, the roadway and walkway were undermined and ultimately closed, and the utility lines broken. Moving the water and sewer lines back to Portola Drive was an expensive proposition and a factor that had to be carefully weighed.

Options for protection included armoring only the base of the bluff, only the upper portion of the bluff, or armoring the entire bluff from top to bottom. Groins were also considered as a way to trap littoral drift and form a protective beach. While a wider beach had clear benefits, the accumulating sand would cover a portion of the rocky intertidal zone, which has some biological impacts. Although a wider beach would reduce wave activity at the base of the bluff, the beach would likely narrow or disappear altogether during the winter months, which would reduce the level of protection when it was needed most.

The County Redevelopment Agency and Department of Public Works held a number of meetings where all those interested had the opportunity to look at the designs being considered as well as their corresponding environmental impacts, ask questions of the county staff, the consulting geologists and engineers, and provide input. The surfing community and the residents of the Pleasure Point area were very involved in the process. The comprehensive Environmental Impact Report (EIR) was written and revised several times in response to questions raised by the public and by Coastal Commission staff. The level of public participation and opportunities for involvement and review provided by Redevelopment Agency staff played an important role in the ultimate project and it's acceptance.

While the Coastal Commission today is not generally in favor of additional coastal armor, the perceived bluff-top public benefits of the project, including better pedestrian and bicycle access, improved parking, and maintaining East Cliff Drive for vehicles, were all seen as positive elements. The protection and enhancement of public coastal access, including new and replacement coastal access stairways, were additional benefits. The proposed design, a soil-nail wall constructed to look as much as possible like the actual bluff materials, was also a major improvement over many of the existing concrete seawalls, rip-rap and other materials used along East Cliff Drive.

This site, for all of the reasons discussed in earlier columns, was a challenge from beginning to end, with many different groups and individuals understandably interested, concerned and involved. From my own experience in similar projects of this sort, I believe it is a very successful example of what can happen when the public agency is open with their plans and intent, that the public takes the time to get involved, attend meetings, listen and speak, and all parties remain open-minded and willing to compromise. All of the benefits envisioned at the outset have been realized and it has become part of this unique neighborhood.

Part III: Coastal Erosion, Protection & Shoreline Change

Article 11

Coastal cliffs and Rolling Rocks

Fallen rocks at Waddell Bluffs, Santa Cruz County, 1978.
(© 1978 Gary Griggs)

Watch for falling rocks! Do you ever notice that sign just as you cross the bridge over Waddell Creek and start along the stretch of Highway 1 below those steep cliffs? I've never been exactly sure what a driver is supposed to do if they do in fact see a large falling rock. The death of a passenger in a tow truck in 1976 from a several hundred pound boulder, which rolled down the slope and through the front window, led to a lawsuit and some major changes in how CalTrans dealt with falling rocks along this old sea cliff.

The sections of Highway 1 that cross Devil's Slide, Waddell Bluffs, Big Sur and parts of the Malibu coast were all hacked out of steep and unstable bedrock or talus deposits. While they provide an up close and often exciting view from the family station wagon, they also come with some warning signs and definite hazards.

As described in an earlier column, the 350-foot high cliffs north of Waddell Creek formerly plunged into the sea and the rocks that regularly fell, slid and rolled down the cliff came harmlessly to rest along the shoreline as a wide talus slope. This begs the question: are there geologic hazards without people?

The Ocean Shore Railroad engineers couldn't figure out how to get tracks safely across the base of the cliffs. Early travelers, whether on horseback, stagecoach, or Stanley Steamer, waited for low tide and then raced across the beach. From an 1849 manuscript provided by my traveling companion Sandy Lydon, one can get the full picture of what this adventure was like 160 year ago.

Justo Veytia, a Mexican citizen, set out on horseback for San Francisco via the North Coast in November of 1849. After passing the landslide at present-day Waddell, he wrote: "Two days of this expedition were the most difficult. The second day on the road one has to travel along the beach very close to the water and this can only be done when the tide is low. The day we passed the sea was quite choppy. Neither Arana nor I knew the road so when we went onto the beach we figured it was all right because when a very big wave came up, it only reached the horses' hooves. So we rode on about 300 varas (about 300 yards), experiencing two very bad spots because of some rocks, when the very rough sea began to wash over us up to the pommel of our saddles. We didn't deliberate in making a decision - to go back was clearly dangerous because the rocks were now under water and we couldn't see the openings between them, so we resolved to continue forward to look for some pass where we could go up, for the waves had us pinned against a fairly high cliff. We went on walking for about 200 varas until we found a foot path to ascend and as soon as we were safe we undressed completely to put our clothes to dry because the waves had knocked us down three times, horses and all, so we had to dismount and pull them forcibly."

During construction of Highway 1 along the bluffs in the mid-1940s, substantial cutting and talus removal took place over about a mile of coastline. Fifty to 120 feet of horizontal cutting into the loose hill slope for a roadbed steepened the lower slopes substantially and required the removal of about a million cubic yards of material. Not surprisingly, the rocks continued to fall from the slopes, only now instead of rolling out onto the beach, they rolled towards Highway 1. Winter rainfall, as well as wetting and drying, and heating and cooling, all weakened the intensively fractured mudstone and siltstone cliffs.

The California Department of Highways realized at the time that rocks would continue to fall and be a hazard to the 280,000 vehicles they originally estimated would traverse this section of roadway each year on their way up or down the coast. Both a barrier wall and a trench were originally considered, but they felt that a barrier wall would be costly, aesthetically unacceptable and only to be used as a last resort. A trench to catch the rocks could be cleaned out periodically, or as required, with little or no inconvenience to motorists. Would it work? Turns out it works if you keep the trench clean, but CalTrans failed to maintain the trench, which led to the fatality.

Article 12

The Ever Changing Coast

Natural Bridges State Beach, c. 1900
(courtesy: Special Collections, University Library, UC Santa Cruz)

If there is one thing that we can all agree on (and there are probably many things we could agree on), it would be that the coastline is never the same from one day to the next. It doesn't take more than two walks along any beach, or along West Cliff or East Cliff, to recognize that where and how the waves are breaking, or how the beach itself looks, is never exactly the same. Trust me, you can count on it. I think this is one of those things that makes the coast interesting and draws us back, day after day, year after year.

It was initially a collection of old photographs of the Santa Cruz coastline from the late 1800s and the early 1900s that led

Part III: Coastal Erosion, Protection & Shoreline Change

to a book we did several years ago ("The Santa Cruz Coast-Then and Now"). We tried to find the exact location where those early photographers stood in order to take modern pictures and capture a century or more of change. In some cases, we discovered that we couldn't stand in the place where the photographer stood a hundred years ago because the spot was gone, lost to the storms and waves.

There are places along the Santa Cruz coast that have undergone some remarkable changes over the past century, and others that have changed surprisingly little. It's a constant battle between the energy of the waves and the strength and weakness of the rocks, and the waves eventually win.

Natural Bridges is a good example of the first, lots of change. Even its name keeps changing. It was originally known as Moore's Beach, then Hall's Beach and later Swanton Beach, before becoming Natural Bridges. To be perfectly honest today, it should be called Natural Bridge, as there is only a single arch remaining.

Why Moore's Beach? Eli Moore arrived in Santa Cruz from Missouri in 1847 with his wife and five children. He bought a ranch that extended from Empire Grade to the coast and through which Moore Creek flows. The creek passes through what is now Antonelli Pond, but which was then known as Moore Creek Lake, and then on to "Moore's Beach" where the creek entered the ocean. I don't think anyone really knows whether Eli Moore named all of these features after himself, or whether they simply came into common use because he was the landowner. Only Moore Creek has held up over the years, however.

When Richard Harrison Hall came from Vermont to Santa Cruz in 1853 he bought 300 acres, including the old Natural Bridges Dairy, out on what was then called the Cliff Road (the Moore property?). The deed included "three natural bridges" and the beach became known as Hall's Beach.

One of the area's early entrepreneurs and developers, Fred Swanton, formed the Swanton Investment Company in 1908 to develop a subdivision along West Cliff Drive next to Natural Bridges known as Swanton Beach Park. With the hope of luring buyers from the sweltering heat of the Central Valley, he named the streets after cities in the Valley (Modesto, Chico, Auburn, Sacramento, Merced and Coalinga). The street names didn't really seem to attract many buyers, however. When people realized that this area was often windy and cold, they lost interest and the development ultimately failed. As recently as 1970, nearly all of the lots along Auburn, Chico and Swanton Boulevard were still empty. In 1933, Swanton deeded 54 acres of the property, known at the time as Swanton Beach or Swanton Beach Park to the state.

The North Pacific waves have taken their toll on the cliff here over the years, however, with one bridge collapsing around 1905, and a second failing in the storms of 1980, leaving us today with Natural Bridge State Beach. In time, sadly, that last bridge will also fail.

But the coastline doesn't change uniformly, even over short distances. Two-thousand feet west of Natural Bridges, directly in front of the Seymour Center at Long Marine Lab, we have good evidence that the cliff hasn't retreated a foot in 90 years. On October 1, 1924, the La Feliz, a vessel headed towards San Francisco from Monterey with a cargo of canned sardines, ended up stuck on a rocky ledge below the cliffs. Local residents came out at night to rescue the crew. The mast was removed, leaned against the cliff, and used with a block and tackle to recover the cargo from the ship. Remains of the ship's drive shaft can still be seen at a low tide on the beach below the cliffs.

But the most surprising part of the story is that mast is still leaning up against the cliff, 90 years later. The resistant mudstone platform that the La Feliz was grounded on has acted as a natural breakwater and protected the cliffs and the mast from heavy wave attack.

Sixteen years ago, I stood in front of the California Coastal Commission, asking for a permit so that we could build the Seymour Center. The chair of the Commission at that time, Sara Wan, said to me "I don't want you coming back here in 30 or 40 years asking for a permit for a seawall to protect your building". I told her the cliff hadn't eroded a foot in 75 years and I promised her that I wouldn't be coming back in for a seawall permit.

On one of our beautiful fall days, bike, walk or drive out to the Seymour Marine Discovery Center and you can still see the mast of the La Feliz leaning up against the cliff. While it may look at first glance like a telephone pole, it is the mast of that shipwreck from 1924, still holding on after 90 years.

Article 13

Coastal Promontories

Illustration of the arch
at the end of Woodrow Avenue, Santa Cruz
(© Jim Phillips)

To a geologist, the entire Earth is a giant jigsaw puzzle. Many of us who study the Earth spend virtually our entire careers looking for clues to the formation or evolution of some piece of the planet, a mountain range, a river valley, an ocean basin, or a stretch of coastline. In my case, the last 46 years has been pretty much focused on trying to understand the coast of California. It's a messy piece of real estate, and its always changing, which further complicates sorting out the pieces and reconstructing the history.

My last column focused on West Cliff and the process of trying to figure out where an old arch stood from today's perspective. I asked readers to let me know what they thought. A local legend, surf and skateboard artist, Jim Phillips, kindly sent me a photograph he had taken in the mid-1970s, which still showed Arch Rock out near the end of Woodrow. Jim had also sent a nice drawing he had made of the arch, which was published in the old Santa Cruz Weekly.

So I went back out to the end of Woodrow, where the old Gazebo once stood, with Jim's photograph in hand to try and see if I could relocate the site of the arch. I was surprised that even with a photograph that was only 40 years old, I still wasn't exactly sure where that old arch had been, but I now have a much better idea.

But then West Cliff has gone through a lot of storms since the 1970s. Literally, millions of waves have battered the cliffs, and taken a lot of rock with them. Despite the endless pounding by the waves, there are some areas that put up a lot more resistance than others, Lighthouse Point, San Lorenzo Point at the river mouth, and Soquel or Pleasure Point are three good examples.

Every point or headland that juts out into the waves is an example of rock strength persisting over wave attack, at least for hundreds of years. The oldest photograph we have come across shows Point Santa Cruz or Lighthouse Point looking about the same in 1876 as it does today.

Seal rock has survived at least as long. There used to be two seal rocks, but one succumbed to the waves sometime between 1956 and 1963.

A logical question I've asked myself many times is, why? Why does Lighthouse Point, San Lorenzo Point or Pleasure Point manage to survive through all the Pacific Ocean can throw at it for centuries, while the rocks on either side have eroded or collapsed?

There is a simple answer, but it may not be very seem particularly earthshaking. The rocks in these areas, and at Seal Rock, are harder or more resistant to wave attack than the rocks to either side.

Most likely it is the degree of cementation or consolidation of the rocks at each of these points that provide their strength. But structural weaknesses also play a role. Joints, which are cracks or fractures produced by years of stress that these rocks have been through, provide zones of weakness that waves can attack like a wedge.

A careful observer can recognize these joint patterns in the rocks on the west side of Lighthouse Point, and along much of West Cliff. They allow the waves to focus their energy and, over time, give the regular geometric shape to much of our local coastline, and also produce the caves that indent and undercut the bluffs along West Cliff Drive. One of these collapsed again during the winter of 2014 and produced a blowhole just east of the end of Woodrow, which the city has now patched up again with concrete.

Arch at end of Woodrow Avenue, Santa Cruz, c. 1970.
(© Jim Phillips)

Waves and a rising sea level will eventually win the battle, but the Purisima Formation bedrock that makes up these protruding points is doing pretty well. This is the same formation that underlies Opal Cliffs and Depot Hill; but the sandstone and mudstone there is more failure prone and retreats at a foot or so each year on average. Seawalls and riprap have temporarily halted or slowed this in many places, however.

Article 14

Untrained River Mouths

Aptos Creek with old timber training wall, Santa Cruz County.
(© 2012 D Shrestha Ross)

River or creeks don't always go where we would like. The residents of The Island in Rio Del Mar, just east of the mouth of Aptos Creek, have become acutely aware of this over the past two winters. The Santa Cruz Main Beach has also had its problems with the path of the San Lorenzo River as it crosses the beach, particularly when it decides to head west and creates a big pond along the back beach. Most beach visitors would rather put their towel on nice dry sand rather than in the water.

As our central coast rivers and streams meander across the beach, heading towards the sea, the littoral drift of sand along the shoreline influences their pathway. From Half Moon Bay all the way to Moss Landing, waves from the northwest generally drive the beach sand downcoast, south or east depending upon just where you are.

This littoral transport of sand acts like a huge bulldozer on the beach, moving about 300,000 cubic yard a year, or 30,000 dump truck loads, along the shoreline. Most of the time, this persistent flow of sand deflects or pushes the stream mouths downcoast. Waddell Creek, Scott Creek, Soquel Creek, and the Pajaro River are all normally deflected south or eastward. The Pajaro River often flows as far as a mile south behind a sand bar before it finally reaches the ocean.

A careful observer or long-term beach visitor will notice that this isn't always the case, however. During the winter months, when these streams are flowing at high velocities, they usually have enough power to blast their way through any sand flow or sand bar and enter the ocean directly.

There are also periods when stream flow is very low and waves may approach from the south and move sand back upcoast. Under these conditions the same streams may be deflected upcoast or to the north. There are times when Scott Creek, for example, hugs the north side of its beach. The Pajaro Dunes development was built right at the north side of the Pajaro River mouth and a timber bulkhead was constructed to keep the river from eating into the development should it decide to suddenly flow northward along the shoreline.

What about the other streams? The San Lorenzo River is trapped up against San Lorenzo Point so it really can't be deflected downcoast. Throughout most of its human history, it has behaved and flowed quietly seaward along this narrow rocky peninsula. During dry summers, which we may be seeing more and more often in the years ahead, the littoral drift of sand often forms a dam across the trickle of water. The San Lorenzo backs up, and over time, will form a pond that may be over two miles long and stretch all the way to the Highway 1 Bridge. As the elevation of the pond slowly rises, the water will follow the lowest path, which may be west in front of the Boardwalk, leaving much of the city's most popular beach underwater.

The same process often takes place in the late summer and early fall in Rio Del Mar. Aptos Creek is reduced to a trickle, the littoral sand bulldozer forms a dam and the creek backs up, creating a pond extending for hundreds of feet along the back beach, often extending both up and down coast.

The residents of The Island, southeast of the mouth of Aptos Creek, have a more serious problem to contend with. The natural tendency here has always been for the mouth of the creek to head east along the beach, pushed along by the sand moving down coast.

This was recognized decades ago, before the first development took place along the beach. Originally, the area along Aptos Creek between the bluffs, now known as the Rio Del Mar flats, was a big lagoon. During the summer months, when sand dammed the creek mouth, the water backed up and flooded most of this low-lying marshy area. When the creek ran through to the ocean, it flowed up against the eastern bluff, pushed that direction by sand flow coming downcoast from Santa Cruz and points north.

These natural conditions were all changed in 1926 when the Peninsula Properties Company channeled the creek with concrete walls and pushed it up against the opposite or west side of the lagoon. With the creek now constrained, fill from the eastern bluff and the cutting of Cliff Drive up the bluff, was used to bury the lagoon with about seven feet of fill. A subdivision, the Rio del Mar flats, was then laid out and lots sold.

In 1928 the company built a concrete dam across the mouth of the creek between the two concrete retaining walls, creating what was advertised as the largest fresh water swimming pool in the world! Two years later, however, during the winter of 1930-31, a storm destroyed the dam and eliminated the world's largest and shortest-lived fresh water swimming pool.

At the same time the creek was channelized, a timber training wall was constructed on the east side of the creek as is crossed the beach in an effort to keep it from heading downcoast. The wall was a double row of pilings that extended about 150 feet onto the beach and is still in place today, although a little worn over the subsequent 86 years. There is a gap between the two rows of piling that may have been initially filled with rock, but rock also has been added along the east side in recent years.

The problem that downcoast beach homeowners have been dealing with is the desire of Aptos Creek to flow where it has always flowed, east or downcoast. It usually turns east as soon as it gets to the end of the bulkhead. In the winter months the creek has been eroding a channel along the beach directly in front of the homes on The Island. The 25 homes here are built on the beach seaward of Beach Drive, so they are an island of sorts. Although their homes were built literally on beach sand, owners are understandably concerned, for both their protective riprap being undermined, and ultimately their foundations being damaged.

I am appreciative of my good friend and local historian, Sandy Lydon, for providing the information on the history of the development of the Rio del Mar flats.

Article 15

Meandering River Mouths

Littoral drift, driven by the dominant waves from the northwest, pushes the sand southward along much of the California coast and tends to divert the mouths of our coastal streams southward as well. This isn't the case all year long, however, nor does this happen along the state's entire coastline.

Littoral transport along portions of the coast north of Cape Mendocino, in southern Monterey Bay, and between the Mexican border and San Diego tends to move beach sand northward for much of each year.

Part of the explanation for this difference is the direction of wave approach in some areas, but more important is what happens to the direction of wave motion as the swells advance towards the coast. Waves are refracted or bend as they enter shallow water, and the degree and direction of this bending is related to the changes in water depth or the bottom contours.

Swells typically approach Steamer Lane from the northwest. As the underwater portion of the wave begins to encounter the bottom, that section of the wave will start to slow down. The entire wave front will then progressively bend or wrap around Lighthouse Point and break as each section enters shallower water. Waves will gradually break eastward from the point, making for the well-known surf break that can extend all the way to Cowell Beach during large swells.

Wave refraction along the northern portion of Monterey Bay drives sand southward towards Moss Landing. This south flowing littoral drift of sand would eventually fill the entrance to the Santa Cruz Small Craft Harbor if we didn't dredge out the 250,000 to

Part III: Coastal Erosion, Protection & Shoreline Change 171

300,000 cubic yards of sand that the waves transport into the entrance channel each year.

The harbor at Monterey is rarely dredged, however. Between Moss Landing and Monterey, instead of being transported south and into Monterey harbor, sand more often moves northward, carried by waves refracted around the Monterey Peninsula and into the southern end of the bay.

As a result, instead of being diverted southward like the other streams entering the ocean along the central coast, the Salinas River was diverted north for hundreds of years. One of the oldest maps around, drawn in 1854, shows the Salinas River flowing north for about six miles from its present entrance. It was separated from the ocean by a narrow dune covered sand spit to a discharge point about a mile north of the present mouth to Moss Landing Harbor.

During the 1800's, the spit and dunes separating the Salinas River from the bay were breached on occasion, not only by the river, but also by ocean waves: "A natural dam has been formed across the Salinas River near Moss Landing by cutting of the surf through the sand hills separating the river from the ocean" (from the Salinas City Index, March 7, 1878).

This northerly path of the river was maintained until about 1910, when through either natural breaching, or perhaps assisted by local farmers to reduce flooding along the lower course of the river, the present mouth was created with a straight shot across the sand spit to the ocean. A map from 1910 depicts the Salinas River still heading north to its ancient mouth north of present day Moss Landing, although the river flow may well have already been diverted. Today, at the entrance to the old river channel there is a floodgate, which can be opened to let river water flow into an irrigation ditch. This is all that exists of the river's former northerly course.

While writings up until about 1850 give the present day Salinas River the names Valle de Monterey, Rio de Monterey, Monterey River and a few others, the 1854 map labels it as the Salinas River for the first time. There have been salt flats near the entrance to

Elkhorn Slough for years, and they are actually designated on the 1910 map. The Salinas River name was derived from the Spanish word salina, for salt flats or ponds. During the Mexican and Spanish eras, the two local sources of salt were either Point Lobos or the ponds at Moss Landing near the old Salinas River mouth.

Humans intervened here in 1946 when the US Army Corps of Engineers constructed a stable channel by dredging the present entrance to Moss Landing Harbor through the sand spit directly opposite the mouth of Elkhorn Slough. The entrance channel was stabilized with two massive rock jetties and the old Salinas River path to the north was abandoned.

Article 16

A Castle on the Beach

Seabright Beach, c. 1931.
(courtesy: Special Collections, University Library, UC Santa Cruz)

Seabright is probably the widest beach in the city of Santa Cruz today. However, it wasn't always that way. Before the west jetty of the harbor was constructed in the early 1960s, even the summer beach here was quite narrow. Most visitors parked their blankets and umbrellas in the cove directly across from the Natural History Museum, as this was one of the few areas where dry sand could be found.

Old photographs and postcards often depict a narrow fringe of sand against the bluff in the summer months, a much different beach than residents and visitors enjoy today. Without a protective beach, winter waves crashed against the bluffs and took their toll on the cliff and East Cliff Drive as well. That hasn't happened for the past 50 years, however.

Seabright has long been a unique neighborhood with a personality and character that has survived for well over a century. It was considered to be out in the country by Santa Cruz standards when it was first developed as a seaside resort in the 1880s. F.N. Mott bought 12 acres, laid out streets and sold lots.

Named after Seabright, New Jersey, the area soon had its own Post Office (now occupied by La Posta Restaurant), as well as a small station on the Southern Pacific Railroad line between Santa Cruz and Watsonville. Seabright Avenue was formerly known as Railroad Street.

Another local pioneering family, the Pilkingtons, also left its mark. Pilkington Gulch is a small intermittent stream running from the general area of Seabright Avenue and Woods Street to the Seabright Cove at the foot of Mott Street.

Thomas Pilkington, who was born in England in 1815, migrated to Mexico where he established a cloth printing company in 1841. He served in the American army during the Mexican war, and then moved to California during the Gold Rush of 1849.

He ended up in Santa Cruz in 1853 where he became one of the early settlers in the Seabright area, obtaining the land between the San Lorenzo River and Twin Lakes and stretching half a mile inland from the bay to Pine Street. This area was hayfields at the time and Pilkington bought squatter's rights after securing U.S. patents to the land.

For years, a rather makeshift footbridge over the San Lorenzo River was the main route into Santa Cruz. Each winter it was removed to keep the river from washing it away, and Seabright residents had to walk across the railroad bridge, considered dangerous at the time, as there was no pedestrian walkway as there is today.

The only road to Santa Cruz from Seabright in those early days was a narrow lane along the edge of the bluff, which went through the Pilkington property. As the Seabright population grew, the residents wanted a more direct route to town so Thomas Pilkington donated a strip of land along the bluff for what would later become East Cliff Drive.

In the 1880s, Pilkington also built one of the first summer camps in the area, Camp Alhambra. One of his two sons, T.B. Pilkington, subdivided the family property in 1891 and named the streets, Pilkington, Alhambra and Brook.

Thomas Pilkington's other son, James, built the Seabright bathhouse in 1899-1900 on the the west edge of the Seabright Beach Cove. The bathhouse was built to look like a castle and is the source of the original name, Castle Beach. It is unknown whether there was any connection between the 13th century Moorish castle and fortress in southern Spain, the Alhambra, and the castle on Seabright Beach and Alhambra Avenue.

In 1918, Conrad Scholl and his son Louis took over management of the bathhouse, and in later years added a dining room and renamed it the Scholl-Mar Castle. Another era for the castle began in the 1940s, when it became a restaurant named Casa del Mar. In the 1950s and early 1960s the castle saw new life as an art gallery.

Throughout its history, prior to the construction of the Santa Cruz Small Craft Harbor and the subsequent widening of Seabright Beach, the waves attacked the bluff, including the castle, nearly every winter. In March 1967, the Castle was demolished, leaving behind a very wide beach with a curious name and an interesting history.

Article 17

Walls Around our Coastal Cities?

The Thames River Barrier, 2004.
(© Andy Roberts, licensed under CC BY-SA 2.0 via Wikipedia Commons)

Hurricane Sandy has come and mostly gone, but the devastation left behind will take years to clean up and repair. The final damage toll was over $68 million for New York and New Jersey alone. Requests for federal disaster aid from FEMA (lots) started before the storm had even passed, and the inevitable Monday morning quarterbacks immediately began devising all of the possible solutions for protecting the coastal areas of New Jersey, New York City and Long Island in the future.

The total damage figures are huge, but the meteorological figures are in some ways, more alarming. Sandy produced the highest storm surges ever to hit New York City, nearly four feet higher than the previous highest water levels during Hurricane Donna in 1960.

Part III: Coastal Erosion, Protection & Shoreline Change

New York City is particularly exposed to the sea, with water coming in through the Hudson and East rivers, and with a meandering total of hundreds of miles of exposed shoreline, much of it just a foot or two above sea level. Sea level is also now about a foot higher here than it was in 1900.

You might think that New York City would have thought about these things many decades ago, but sea-level rise and storms probably weren't as pressing in the 17th century when the city was first laid out. The early settlers were Dutch, who knew a thing or two about living very close to sea level. But the founding fathers must have felt that a few feet of freeboard was enough, or simply had other issues to deal with.

To add insult to injury, sea level along the Atlantic coast is likely to rise 3 to 4 feet or more by 2100, as the oceans continue to warm and ice sheets continue to retreat and melt. Sadly, the worst isn't over and it's not likely to get better any time soon. So despite the energetic calls to re-build and armor the shoreline against future disasters, there is wisdom in pausing for contemplation before jumping to employ the first band-aid from the bandwagon of ideas to come along.

There are a lot of greenhouse gases already in the atmosphere that we can't get back, and we don't have a switch in place to turn off what we are putting in every day now. Emissions continue to rise. In 2012, the Earth's 7 billion people dumped 39 million tons of carbon dioxide in the atmosphere every day. Every day! Where is all this coming from? Well, 91% comes from burning fossils fuels and the manufacturing of cement, while the remaining 9% comes from land use activities (burning forests, and agriculture). And the overall emissions continue to rise each year.

Although China has now surpassed the USA in total carbon dioxide emissions, there really isn't much comfort in this statistic as we all share the same atmosphere. On the positive side, carbon dioxide emissions from the USA and the European Union have dropped in recent years, but on the other side of the fence, emissions from

China and India are increasing. Off all those 39 billions of tons of carbon dioxide we emit globally each year, 50% remains in the atmosphere contributing to the warming we are experiencing, and 26% is taken up by land vegetation on the land. The remaining 24% ends up in the ocean, being taken up at the rate of just over one million tons every hour. This has consequences as well, primarily by making the oceans more acidic. But that's another story.

So what are the planners, designers and engineers coming up with for New York City? There is the $6 billion system of floodgates that has been proposed, using Venice, The Netherlands, and London as examples. This approach raises some important questions about who or what gets to be protected and which neighborhoods are on their own

Ten years ago, a research team from the State University of New York at Stonybrook, working with the state's Environmental Protection Agency, put together a complex plan of levees, seawalls and floodgates for protecting most of New York City from storm surges. The estimated price tag of $10 billion at that time brought that proposal to a screeching halt.

Some are calling for a mix of hard structures, whether seawalls or storm barriers of some sort, combined with "soft strategies", like constructing urban wetlands, tidal marshes, even artificial reefs intended to nurture oysters, which has been referred to as a blending of urbanism and ecology.

The hope here is that these engineered green spaces would absorb and reduce the force of incoming water, thereby protecting the shoreline. Porous streets of concrete have been proposed to soak up excess water like giant sponges, while other new streets would be designed to drain the surging water back into the harbor.

If Manhattan was a sparsely inhabited rural area, these ideas might be considered practical, but this is New York City - protecting it with oyster reefs and some wetlands? When everything - streets, subways, and underground parking garages - are all completely underwater with a storm surge of nearly 14 feet, porous concrete is not going to soak up the excess water, or drain the water back into the harbor. There is a need to be creative, but also a need to be realistic and fully consider the forces and water levels we are seeking to control, and the fact that sea level is going to continue to rise.

Its also important to come to grips with the reality that New York City isn't the only large city in the United States (forget the rest of the world for a moment) that lies very close to sea level and is exposed to hurricanes, extremely high tides and storm surges, or a rising sea. There are a few others to consider, in no particular order: Miami, Newark, New Orleans, Tampa, Boston, Wilmington, Virginia Beach, Charleston, Galveston, and on the west coast, Long Beach and parts of many of the low-lying communities surrounding San Francisco Bay, including San Francisco, Oakland, and San Jose.

Dr. John Holdren, Director of the White House Office of Science and Technology and former Professor of Environmental Policy at Harvard University, has said "With climate change we basically have three choices: mitigation, adaptation and suffering. We are going to do some of each. The question is what the mix will be. The more mitigation we do, the less adaptation and suffering there will be". We need to start making some very serious commitments to significantly reducing the emissions of greenhouse gases if we are to avoid a future that is not going to be pleasant for any of us.

PART IV

BEACHES

Article 1

Why Are Our Beaches So Fine?

Pebble Beach pebbles.
(© 2008 Gary Griggs)

To many of us who live along the coast, beaches define California. It's our image of the Golden State. Whether the wide, palm tree-lined beaches of Santa Monica, the boardwalk-backed Main Beach in Santa Cruz, or the rugged, rocky beaches of Mendocino, there is a beach for almost everyone.

If we divided up California's 1,100 miles of coastline by the state's 38 million residents, however, we would each have less than two inches of shoreline. Being more realistic, only about 300 miles of our state's coast actually consists of accessible beaches, so we each really have about a half an inch of oceanfront to enjoy.

To stress the shoreline a little more, our share is reduced further because we have to share this with some of the state's 32 million annual visitors. While California's population has doubled over the past 40 years, our beaches haven't gotten any larger, and in some places, they've actually gotten smaller.

Have you ever stopped to ask yourself why we have sandy beaches at all? Why not just steep rocky cliffs? Or instead of soft white sand, what if beaches consisted of round cobbles the size of golf balls or baseballs? While we do have some beaches like this, can you imagine jogging or playing volleyball on golf balls? It would be painful to say the least, and I think its fair to say that most people would probably be doing something else on warm summer days instead of going to the beach.

Beaches are sort of a geological coincidence. Most sand on California's beaches comes from our rivers and streams. Weathering of the rocks in a watershed or drainage basin through heating and cooling, wetting and drying, and freezing and thawing, gradually breaks down the bedrock into smaller fragments. These are carried down slope by gravity, rainfall, and runoff, until they reach a stream. High winter flows in creeks and rivers gradually move the rocks downstream, breaking them up, sorting them out, and rounding and smoothing them along the way. The large boulders are left way upstream in places like Boulder Creek; the pebbles are carried farther downstream to places like Felton, and it's often only the sand and finer-grained silt and clay that make it to the coast. The sand is deposited on the shoreline, but waves along most coasts are energetic enough to keep the silt and clay in suspension. Ultimately they carried offshore, often tens or hundreds of miles before finally settling out onto the seafloor.

Breaking waves along many coasts are most effective at transporting and sorting sand. Gravel and cobbles are usually too heavy for most waves to move very far, and silt and clay are so light that they are transported off the beach. You just don't find ocean beaches made of silt or clay.

Streams tend to deliver sand-sized material to the shoreline and waves prefer to sort and move sand. So we end up with this ribbon of soft sand along most shorelines, which the waves have sorted, smoothed and rounded for our pleasure. Soft sandy beaches are something we often take for granted. Santa Cruz would be a very different place if the shoreline from Cowell to the harbor were covered with rocks instead of sand. More to come…

ARTICLE 2

Beaches-Here Today, Gone Tomorrow

Our California beaches usually look stable and permanent when we visit during the summer months. And they generally are pretty stable for much of the year, but the sand rarely sits still for very long. Waves are constantly at work, pushing the sand back and forth across the beach face. Wind may also blow the sand inland from the back beach and form dune fields like we have in central and southern Monterey Bay.

Seasonal beach changes are probably most evident to those of you who walk along the beach year round. About November or so each year, the first serious storm waves usually begin to hit the shoreline, and it is these larger, more energetic waves that stir up the sand and begin to carry it offshore. The wide summer berm, or the area where you parked your towel or played volleyball in July and August, is gradually removed. With narrow beaches like Its Beach, or those along East Cliff, the dry sand may be completely gone from December to March during most winters. The wider beaches, Cowell, Main Beach and Seabright, will usually still have some dry beach remaining for the winter die-hards. In the big El Niño winters such as 1978, 1983 and 1998, however, even these beaches will narrow and erode, with the wide summer beach typically replaced with logs, trees and other debris from the rain-swollen rivers- at least during rainy winters.

The sand that is moved offshore each winter tends to align itself into a set of sand bars and troughs that are parallel to the beach. This is the shoreline's way of readjusting itself to the more energetic winter waves. Because the depth where the waves break is determined by the height of the waves, the presence of these sand bars

Fall at Its Beach, 1997.
(© 1997 Kristen Brown)

causes the larger waves to break farther offshore. This dissipates more of the breaking wave's energy offshore, which reduces the amount of energy expended on the beach, a natural shock absorber for the shoreline. Even with these buffers, however, winter waves are still the ones that do the most damage to the coastline, and to anything we've built within reach of the waves, whether houses, restaurants, streets, water or sewer lines, parking lots or bike paths. And as sea level continues to rise in the decades ahead, the waves will gradually reach farther inland and claim more of our coastline.

By late spring, the wave climate has usually changed and the waves are now lower and less energetic. They begin to transport the sand from the offshore sand bars landward, grain-by-grain, and gradually rebuild the narrow winter beach. By July or August the beach is usually at its maximum width again, just in time for all of the summer tourists.

The balance between a winter and a summer beach, or the wave conditions that cause sand to move either offshore or onshore, is

Winter at Its Beach, 1998.
(© 1998 Kristen Brown)

somewhat delicate, and conditions can change quickly and reverse the transport of sand. Years of observations have shown that wave steepness, which is the ratio between the wave height and the wave length (or the horizontal distance between wave crests), exerts the strongest influence on whether the sand moves onshore to form a wide berm, or offshore to form a sand bar.

Article 3

Beaches - Moving On

Santa Cruz Harbor.
(© 2006 D. Shrestha Ross)

There is nothing very permanent about a beach as the sand grains are only temporary residents. Not only do they move offshore each winter and back onshore each spring, but in most places along the California coast the grains are moving along the shoreline as well. This transit of sand is called littoral drift. Every time a wave breaks it suspends or picks up billions of grains of sand, probably trillions but I've never tried to count them all. That sand is not only moved up and down the beach face, but depending upon the angle at which the waves approach the coast, the sand may also move along the shoreline, up or down coast. Most waves approach the shoreline at some angle, simply because the storms that generated the waves were somewhere far out to sea to the north or to the south, but not usually directly offshore.

The uprush and backwash of each breaking wave will carry sand grains in a zigzag pattern along the beach face, moving them a short distance, perhaps a few inches or a few feet down coast. If you are out swimming or floating in the surf zone for any period of time, you will often notice that you have been carried a short distance down coast from where you started. The sand grains are being carried alongshore by the same water movement.

This transport of sand or littoral drift can be thought of as a river of sand moving parallel to the shoreline. At first glance, this might not seem like a particularly effective mechanism for transporting sand, at least compared to a large river in flood stage. But if you stop and think about it for a minute, typical waves breaking along the Santa Cruz coast may have an average period of 8 seconds. In other words, every 8 seconds another wave will break. Some days it may be 7, 9 or 10 seconds, but 8 seconds is typical. There will be 450 of these 8 second waves breaking on the beach every hour or 10,800 every day! That's actually a lot of energy to move sand around.

If each breaking wave moves individual sand grains only 1 inch along the beach face, this amounts to 900 feet of transport along the shoreline every day. Because the more energetic winter waves usually approach the California coast from the northwest, the littoral drift of sand along most of the state's beaches is from north to south. So sand in Santa Cruz moves from Cowell to Main Beach and then on to Seabright over the course of hours or days.

How much sand can be transported as littoral drift? Well, it's a very difficult thing to measure. The best indicators we have along the California coast turn out to be where we have built harbors. Santa Cruz is a good example. After the sand has reached the south end of Seabright or Castle Beach, it is carried by waves around the jetty and into the harbor entrance where it must be dredged out in order to keep the channel open. In recent years, the dredge has removed about 250,000 yds^3/yr on average, although this varies somewhat from year to year. If this sand were being transported along the shoreline in dump trucks, rather than by littoral drift, it would require a truck every 20 minutes, 24 hours a day, 365 days a year.

ARTICLE 4

Moving Mountains of Sand
Dredging California's Harbors

Santa Barbara Harbor..
(© 2004 Bruce Perry, Geological Sciences, California State University Long Beach)

Every year the dredge at the Santa Cruz Harbor sucks up about 250,000 cubic yards of sand from the entrance channel and pumps it out onto Twin Lakes Beach where it continues its journey down coast. If it were put in dump trucks, it would fill about 25,000 of them, but the waves move all that sand without any carbon emissions. After being carried along the shoreline of East Cliff, Capitola, New Brighton, Rio Del Mar, Seacliff, Seascape,

Manresa and Sunset beaches, all that sand flows down into the head of Monterey Submarine Canyon, never to be seen again.

The dredging is an endless and thankless task, but we really don't have much choice if want to maintain the harbor. If it's any consolation, a lot of other California coastal harbors share the same challenge - how to deal with all the sand that continues to move into their entrance channels. Santa Barbara harbor dredges a bit more than Santa Cruz, about 300,000 cubic yards annually, while a short distance down coast, Ventura must move about 600,000 cubic yards a year. A few miles farther east, Channel Islands harbor dredges close to a million cubic yards, every year! At about $2/cubic yard, these are costly undertakings. And the problem will never go away and will get gradually more costly as fuel and labor costs increase.

Interestingly, not all of California's harbors have sand problems. Neither Moss Landing nor Monterey harbors do any significant dredging. King Harbor in Redondo Beach doesn't need to move sand, and Dana Point Harbor has never been dredged. Why is that? Why do we have to move 250,000 cubic yards of sand every year out of the Santa Cruz Harbor, and 20 miles away, Moss Landing has a channel that doesn't need to be dredged? Just like real estate, its all about location, location, location.

Sand moves along the shoreline of California within essentially self-contained beach compartments or littoral cells. The sand that moves along Cowell or Main Beach is in a completely different compartment than the sand coming out of the Golden Gate or found along the beaches of Santa Barbara or Santa Monica.

Each cell or compartment consists of 1) sources of beach sand (rivers and streams and some bluff erosion); 2) littoral drift, driven by waves typically coming from the northwest, which move sand southward along most of the California coastline; and 3) sinks, or places where sand leaves the shoreline. In California the major sinks are either sand dunes, such as those along the shoreline of southern Monterey Bay or Pismo Beach, where wind transports the sand inland off the beach; or deep submarine canyons, where sand flows offshore and down slope to the deep-sea floor 10,000 to 12,000 feet below.

Monterey Submarine Canyon is one of the largest in the world, but there are many others along the California coast: Soquel Canyon, Carmel Canyon, Hueneme Canyon, Mugu Canyon, Redondo Canyon and Newport Canyon to name a few. The Santa Cruz beach compartment begins north of Half Moon Bay and terminates at Moss Landing where the head of Monterey Canyon extends almost to the shoreline. After being transported for 90 miles along the shoreline, the sand that began its journey near San Pedro Point, about 15 miles south of the Golden Gate, disappears offshore right at the entrance to Moss Landing harbor. As a result, very little sand enters Moss Landing Harbor so there is almost no dredging needed.

Article 5

Dams-Cutting off our Beach Sand

San Clemente Dam, Carmel River.
(© 2007 Gary Griggs)

In addition to providing most of California's drinking water, the state's rivers and streams also provide about 75% of our beach sand. In order to provide water for agriculture, industry and homes, as well as flood control, recreation and hydroelectric power, the state's rivers were extensively dammed throughout the last century. There are now more than 500 dams impounding rivers and streams that drain directly into the Pacific along our coastline. Since the first one was built in 1866, an average of 3.5 dams have been built each year ever since. In addition to storing water, the reservoirs behind each of these 500 barriers also trap sand that used to be carried to the shoreline.

Dams now withhold sediment from about 16,000 square miles of the state's coastal watersheds and have reduced the flow of sand by 25%, or about 3.6 million cubic yards each year. That's 360,000 dump truck loads annually. The total amount of beach sand now trapped behind all of California's coastal dams totals about 200 million cubic yards, or a line of dump trucks bumper to bumper stretching completely around the world nearly four times. These reductions vary regionally: in northern California, there are still a number of relatively undisturbed rivers, the Smith and Eel rivers, for example, and sand supply has only been reduced by 5%; in central California, 31% of the sand flow has been eliminated by dams; and in southern California, hundreds of water supply lakes, reservoirs, flood control projects, and debris basins have reduced sand supply to the coast by about 50%.

The long-term sustainability of California's beaches depends primarily on the periodic delivery of sand and gravel from coastal rivers and streams. Sand supply to the state's beaches wasn't a concern, however, when most of California's streams were dammed. The benefits of more water, flood control, recreation, and hydroelectric power in some cases, were compelling to farmers, fishermen, and politicians through much of the 20th century. Over the past 25 or so years, however, the side effects of dams, including loss of salmon migration and spawning grounds, as well as beach sand impoundment, have become more and more evident.

The construction of new dams has come to a halt in California and in other states as well, and is actually now going into reverse. A number of dams that formed barriers to fish migration and that were completely full of sediment have been taken down across the country and others are being studied for removal. In California, San Clemente Dam on the Carmel River, Gibralter Dam on the Santa Ynez River, Matilija Dam on the Ventura River, and Rindge

Dam on Malibu Creek are now completely full of sediment and no longer serve any purpose. Additionally, because they were built many decades ago, their seismic safety is now a concern. Combined these 4 dams have trapped 2.8 million dump truck loads of sand that can end up on the downstream beaches. Google "dam removal" to see some of the dam removal projects that have already been accomplished. We can do the same in California and get the sand from these dysfunctional reservoirs onto the beaches where it belongs.

Article 6

Messing Around with Beaches

Santa Monica Beach..
(© 2002 Kenneth & Gabrielle Adelman,
California Coastal Records Project, www.californiacoastline.org)

Hundreds of dams on California's streams have trapped millions of cubic yards of sand that would have been carried to the shoreline under natural conditions and nourished our beaches. Sluices or pipes could have been built beneath these dams to allow sand to pass through, but this wasn't a consideration when these dams were built many decades ago. Those sediment filled dams that no longer serve any purpose are being evaluated for removal, but this has been a complex and time-consuming effort with only a few dam removals thus far. The largest dam removals in the U.S. occurred on the Olympic Peninsula of Washington State in 2012, when two dams on the Elwha River were removed, releasing tons of sediment and allowing the salmon to return to their upstream spawning grounds.

The logical effect of all of the sand trapped behind those dams would be to reduce the size of the beaches, at least those that are downstream or down coast from the dammed rivers. Do we see these narrowed beaches? Well, there are certainly lots of narrow beaches, and many other areas with no beaches at all; but have beaches in California systematically narrowed? Most of the major dammed streams are in southern California, with about 50% of the original sand flow now trapped behind dams, so this is where we should have experienced the greatest beach losses.

Ten years ago, we completed a study of long-term (60-80 years of historic aerial photographs) changes to the beaches of southern California from Santa Barbara to San Diego. Somewhat surprisingly, despite all of the dams, there were no long-term consistent beach erosion trends identified throughout any of the littoral cells or beach compartments including the Santa Barbara, Zuma, Santa Monica, San Pedro and Oceanside littoral cells.

For those relatively natural beaches along this 300-mile stretch of coastline, (those that have not been altered significantly by breakwaters, jetties or groins), analysis of aerial photographs reveals decadal scale narrowing and widening of up to 100 feet related to El Niño and La Niña climatic conditions. Beaches tended to be narrower during El Niño dominated periods when waves tend to be larger and more erosive, and wider during La Niña dominated periods.

Many of the beaches of southern California, however, have been under human influence for decades. The large breakwaters and jetties built to protect harbors and marinas (Santa Barbara, Ventura, Channel Islands, and Oceanside, for example, as well as Santa Cruz) have led to beach widening updrift and erosion downdrift of these structures. Dredging and sand bypassing have usually been necessary to counter these effects. In addition, a number of large coastal construction projects such as dredging out new marinas, large construction projects in the dunes, and river channel clearing, added about 170 million cubic yards of sand to the beaches between Santa Monica and San Diego between 1930 and 1993; that's 2.7 million cubic yards or the equivalent of 270,000 dump trucks a year!

Many beaches were thus artificially widened for decades, but the nourishment from these large public works projects ended some years ago. The El Niño storms from the late 1970s to the late 1990s also temporarily narrowed some beaches. And then there are many beaches, particularly those along the Malibu coast and in northern San Diego County, which have always been narrow. So the disruptions created by large coastal engineering structures and the artificial addition of large amounts of sand, as well as a changing wave climate, may have partly overshadowed the reduction of sand from dam construction.

Article 7

Beach Sand Burglary

Sand Mining on Monterey Bay, Marina.
(© 2005 Kenneth & Gabrielle Adelman,
California Coastal Records Project, www.californiacoastline.org)

In addition to the natural loss of beach sand from wind blowing onshore, sand was historically mined directly from some California beaches and dunes. Three major sand-mining companies removed sand directly from the beach in southern Monterey Bay for about 80 years. The coarse, smooth, rounded, amber colored quartz sand was in great demand for many industrial uses, including water filtration, abrasives, and various industrial coatings. One utility company in San Jose stipulates that utility trenches should be backfilled with this particular sand. As a result, southern Monterey Bay has been the most extensively mined shoreline in the U.S.

Sand mining near the mouth of the Salinas River started in 1906 and eventually expanded to six commercial sites: three at Marina and three at Sand City. Five of these operations used large draglines and huge buckets to mine sand directly from the surf zone. The sixth sand plant is about 2.3 miles south of the Salinas River mouth where sand is extracted by a floating dredge from a man-made pond on the back beach. The plant and piles of sand can be seen as you drive south towards Monterey on Highway 1.

Severe wave erosion of the southern Monterey Bay dunes during the 1980s and early 1990's raised the question of whether the mining of 150,000 to 250,000 cubic yards of sand from the beach every year was the cause of the ongoing shoreline retreat. The sandy bluffs between Marina and Del Monte Beach continued to erode at rates of 2 to over 6 feet a year. Stillwell Hall at Fort Ord, the Monterey Beach Hotel and the Ocean Harbor House condominiums were all threatened and partially undermined by continuing dune erosion. Stillwell Hall was ultimately demolished; a new seawall had to be built to protect the hotel, and two different revetments and now a new seawall have been built to save the condominiums.

Retreat of the shoreline continues. You just can't remove 15,000 to 25,000 dump truck loads of sand each year from a short stretch of shoreline and not expect some response. Waves expend a lot of energy on the southern Monterey Bay beaches; as the miner's draglines lowered the beach level, the waves began to break closer to shore, and expended their energy against the sandy bluffs, which began to erode more rapidly.

Beach sand mining began in 1906, but remained unregulated until 1960. Even then, for proprietary reasons, the sand extraction industry did not have to tell the public how much of our sand they were actually removing each year. In the mid-1980s, after a connection between sand mining and shoreline erosion rates was recognized, all but one of the permits were terminated.

The remaining sand company, CEMEX (one of the world's largest building materials suppliers and cement producers) continues to dredge about 235,000 cubic yards of sand annually from a back beach pond, a rate equivalent to all of the former sand mining operations combined. Despite what appears to be a clear relationship between sand removal volumes and shoreline erosion rates, and the threats and ongoing problems created by that erosion, the mining has fallen outside of the jurisdiction of any permitting agency for over 20 years and continues today. Public beach sand is being removed by a private company and having major impacts, but no state agency has taken any steps to halt the loss of sand and shoreline erosion along the edge of a National Marine Sanctuary.

Article 8

Gold in Beach Sand

Searching for gold on Twin Lakes Beach, c. 1890.
(courtesy: Santa Cruz Public Libraries, CA)

Walking along most Santa Cruz County beaches in the summer months you'll find that the sand looks pretty typical, white or a light tan in color, due to the dominant minerals, quartz and feldspar. Both minerals are durable and persist in sediments for eons. As well as being light in color, both are also fairly low-density minerals. In the winter months, however, a walk on some of the area's beaches will reveal concentrations of black or dark green sand. At the mouth of the Big Sur River you can even find pink or purple sand, colored by the mineral garnet, which is a very hard mineral used for making sand paper.

The black or dark minerals, such as magnetite, ilmenite, and chromite, can be separated from the lighter colored sand by a hand magnet, like you used to do in the sand box as a child - well some of us who played in sand boxes did. These dark minerals contain heavy elements such as iron, titanium, and magnesium, and these denser grains are left behind by the large winter waves as they remove the lighter quartz and feldspar grains. The black sand will often be concentrated in small rills, channels, or in the troughs of the ripples on the beach surface.

The processes that concentrate these heavier minerals on beaches or just offshore are similar to those that left gold behind in the streams of the Sierra foothills. Currents, whether in rivers or driven by waves or tidal fluctuations along the shoreline, will tend to sort out or separate the lighter from the heavier minerals. These concentrations of minerals are known as placer deposits and led to California's Gold Rush. The beaches and shallow offshore waters of Australia are the source for about 95% of the world's rutile, an important titanium mineral, as well as gold, zirconium, tin and chromium bearing minerals. Much of the world's tin comes from similar sand deposits along the coasts of Malaysia, Indonesia and Thailand.

For a short while in the 1920s, the Triumph Steel Company "owned" nearly 2 miles of beach along the northern Monterey Bay shoreline and was mining the black sand, which contained 500 to 1,100 pounds of magnetite per ton of sand. They used a magnetic separator to remove the magnetite and then utilized a furnace to produce a red iron oxide that was used in the manufacture of paint. While this could have happened in the 1920s, mining beach sand and setting up a furnace on the beach would probably not be viewed very favorably by the Coastal Commission today.

The shoreline south of Año Nuevo, and the beaches between Aptos Creek and the Pajaro River are also characterized by seasonal concentrations of black sand. In addition to iron bearing minerals, black sand may contain small amounts of gold, platinum and other rare but heavy metals. A black sand gold rush started along the northern Monterey Bay shoreline near Aptos in the summer of 1860, and by August, miles of beach had been staked off and more than 25 mining claims filed. The digging continued up until the 1880's when one family drilled a tunnel 300 feet into the bluff and for a time were extracting $5 of gold for each ton of sand.

Article 9

The Colors of Beaches

A minute sample of my beach sand collection.
(© 2009 D. Shrestha Ross)

Sandy Lydon and I did a short beach walk adventure one weekend, and took 40 people from the main beach in Carmel along the coastline to Monastery Beach, almost to Point Lobos. This was considerably less demanding than our summer hike completely around Monterey Bay, but equally packed with history and geology. Normally, I try to cover the past 200 million years or so of Earth history on our trips and Sandy covers the last 200 years of human history. It seems like a reasonable distribution of accountability.

One of the most striking things about the main beach in Carmel is the absolutely brilliant white color of the sand. The sand there doesn't seem to move much up or down coast, and as a result, has been worked and reworked many times so that it is almost pure quartz. This is what gives it the distinctive white color. Beaches on the Santa Cruz side of the bay are not quite so white, but on the other hand, our beaches are almost always much warmer and sunnier.

A century ago, the newspapers in Santa Cruz and on the Monterey Peninsula were in a more or less unspoken competition to attract tourists and their dollars. At that time there was an article in the local paper about the differences in weather on opposite sides of the bay. According to the story on the Santa Cruz side of the bay, when children in Monterey or Carmel turned 10 years old their parents would take them to Santa Cruz to see the sun for the first time. I don't know whether this story attracted more tourists, but it made the folks on the north end of the bay feel a little better about their lack of grand hotels and wealthy visitors.

While most Monterey Bay beaches are on the white or tan side of things, sand can be a lot of different colors. While a graduate student at Oregon State University, I had the opportunity to spend a summer studying the coral reefs of Bermuda, about 600 miles off the Atlantic coast. Bermuda advertises its "pink coral sands" as an attraction to draw tourists. And they are pink, but its not coral that gives them this color. It's actually a less familiar organism, a single cell animal known as a foraminifera, which makes a very small pinkish shell. Although not particularly abundant, the small amount of pink in the otherwise white sand gives a distinct color to the beaches of Bermuda.

Fifty miles south of Santa Cruz at the mouth of the Big Sur River, a picturesque walk from Highway 1 through Andrew Molera State Park, you can find purple beach sand. The purple is from a very hard mineral, garnet, which weathers out of metamorphic rocks high in the Santa Lucia Range. Much of the sand paper you use is coated with garnet because it is a hard and abrasive mineral. Wind and waves have concentrated this purple mineral in interesting patterns along the beach for a considerable distance south of the river mouth.

In the winter months you can find black sand along many of Santa Cruz's beaches. Because these dark minerals are heavier than the clear or lighter colored quartz and feldspar that make up the great bulk of most beach sands, the wave run-up and backwash will typically concentrate these darker minerals into distinct patterns on the beach face.

Three thousand miles west in the Hawaiian Islands, there are really only two things to make beach sand out of, either coral and the shells of other tropical organisms, or volcanic rock. As a result, you either see gleaming white beaches or black sand beaches, often with concentrations of a green mineral, olivine, common in the basaltic lava that makes up Hawaii. You may have guessed that I have a fascination with beach sand and I've collected hundreds of different colored beach sands from all over the world, which now are in little glass vials on windowsills and book shelves all over my office and house.

Article 10

Moving Sand Around

Santa Cruz shoreline from the Municipal Wharf to the harbor.
(courtesy: US Geological Survey)

Watching the huge waves breaking along the coast in April 2012 was a clear reminder to all of us of the power of the ocean. Those people who were rescued from the surf during that swell have perhaps an even greater respect for the sea.

Bulldozers spent two weeks or so in mid-March moving sand around on Main Beach in an effort to protect the seawall fronting the Boardwalk by redirecting the San Lorenzo River back to its normal route alongside San Lorenzo Point. Historically, the river has usually maintained a course directly to the ocean. However, when the shoreline here was altered in the early 1960s by the construction of the jetties at the Santa Cruz Harbor, this stretch of shoreline began to slowly be transformed.

Prior to harbor construction, Seabright or Castle Beach had always been quite narrow. In the old days, waves reached the back of the beach each winter, and the sandstone bluffs were actively eroding. A walk along East Cliff Drive today above Seabright, between 1st and 4th Avenues still reveals isolated remnants of the old concrete roadway perched like pedestals thirty feet above the beach.

In 1963, as soon as the west jetty was completed, Seabright Beach began to widen. It continued to widen for the next twenty-five years or so, as littoral drift from upcoast was trapped against the jetty, which acted just like a dam. Where waves previously crashed against the base of the bluff, a sandy beach 300 to 500 feet wide formed, offering year round protection except during periods of high tides and very large waves. It also provided an inviting wide new beach for residents and summer visitors.

As Seabright Beach continued its expansion further seaward, it eventually extended out as far as the end of San Lorenzo Point, and beyond in recent years. The damming effect of the jetty gradually began to impact Main Beach as well. Sand backed up on the beach and widened it. Aerial photographs taken during the summer months over the past 25 years often show a wider sand bar forming at the mouth of the river.

Many of the streams along California's central coast have very low summer flows because of our Mediterranean climate and are typically dammed by sand bars during the summer and fall months. The Salinas, Pajaro and San Lorenzo rivers often experience this phenomenon, as do Aptos and Soquel creeks. These backed up streams form ponds, which provide warm water for kids to play in, a site ripe for bacterial contamination, but also apparently a habitat in places for young fish.

As the level of the pond at the San Lorenzo River mouth increases in elevation as the summer progresses, there is also an increased risk of natural breaching at a low tide when there may be a 6 to 8 foot elevation difference. This can result in a rapid, high velocity and very hazardous draining of all the dammed up water as it flows into the ocean. So for many years, the sand bar that formed at the mouth of the San Lorenzo was artificially breached by the city as a safety precaution.

In the 1980s, however, the California Department of Fish and Game (now Fish and Wildlife) made a decision to eliminate all artificial breaching of the river mouth based on the importance of the pond for fish habitat. The regular breaching was terminated and sadly, there was a tragic accident at the river mouth the next year from an uncontrolled breach. A young woman wading across the river was swept offshore in the fast moving flow and drowned, but was resuscitated although with some permanent medical problems.

At about this same time, the sand bar and beach at the river mouth were gradually getting wider as sand backed up by the harbor jetty continued to extend further to the west. The overall effect of this additional sand accumulation, when combined with the typical low flow of the San Lorenzo in years with minimal rainfall, has been a wider dam across the mouth of the river.

In response, the river in recent years has often taken a different course, flowing west along Main Beach where the beach is often at a lower elevation than the crest of the sand bar. This has led to a large pond on the beach in front of the Boardwalk at times. The entire flow of the river this year went west, along the base of the retaining wall that protects and supports the Boardwalk.

Concern with continued erosion and potential undermining of the wall led to an emergency plan to use bulldozers and a crane to cut open a channel through the sand bar and allow the river to follow its historic course. Rerouting the river also included moving sand from the excavated channel along San Lorenzo Point and from the beach, to build a levee to redirect the river.

The levee was completed on the evening of March 19, with the river being successfully redirected along its historic route directly to the sea. This left a large pond on the back beach directly in front of the Boardwalk. But with the very large waves of this past weekend, all of that sand moved by the dozers to build the levee was all redistributed and smoothed out with a few days of typical winter wave action. There is no trace of the levee left by the next week.

Waves along our shoreline here typically transport about 250,000 cubic yards of sand down coast each year on average. This is equivalent to 25,000 dump truck loads annually, or 68 truck loads every day. It's a very efficient mechanism for transporting sand, as long as it's going where you want it to go.

San Lorenzo River Mouth, Santa Cruz..
(© 2005 Kenneth & Gabrielle Adelman,
California Coastal Records Project, www.californiacoastline.org)

Article 11

Italian Beaches

Monterossa al Mare, Cinqueterra, Italy.
(© 2012 D. Shrestha Ross)

I experienced some coastal culture shock while driving along the Tuscany coast of Italy a week or so after our 3-day June 2012 walk around the shoreline of Monterey Bay. While Main Beach in Santa Cruz can get a bit crowded on most summer weekends, from New Brighton Beach to Monterey most of the bay's shoreline is relatively unpopulated.

Despite the presence of millions of people only an hour or two away, it is easy to find a stretch of Monterey Bay beach with only a handful of people, even on a Saturday in June. There are sections of shoreline where we hiked for an hour or more without ever seeing another person.

Things were a bit different along the Tuscany coast. Italy was hot in June and July, very hot in fact, like many other places in the world that summer. Locals said it was unusually hot. As a result, the Italians were migrating to the beach in very large numbers.

At any access route to the water along the west coast of Italy between Pisa and Rome, there were hundreds of cars parked along both sides of the highway, bumper to bumper. The tiny Smart cars, which are far more common there with gas at $6 to $7/gallon, took advantage of any small gap between cars and parked at right angles to the others, bumper to the curb.

There are few large rivers along the northern Tuscany coast, and as a result, very little sand is delivered to the shoreline and sandy beaches are relatively rare. This doesn't stop the Italians from enjoying the Mediterranean, however. They were descending steep rocky trails to get to the water's edge where they were laying out towels on slabs of rock, large boulders or any other relatively flat surface. And they were doing this anywhere they could find a path to the sea.

Riprap, breakwaters, jetties or slabs of concrete were frequently covered with people. Any reasonably flat surface large enough to sit or lay on worked fine and the earliest arrivals laid claim to the spot, although they might find someone else a few feet away before long.

Where an actual beach of some extent exists, whether sand or gravel, a business arrangement is set up, apparently between an entrepreneur and the local government agency. Virtually every square meter of dry beach is rented out for the day, along with an umbrella and a few beach chairs. An 8 x 10 foot area along with the furniture might go for $20 to $30/day, which includes use of a small changing room. And there are people stuffed right next to you on all sides.

In many of these commercialized beach areas, there is essentially no place for the casual visitor to sit that is not under control of one of the concessions or is not under water. There is really no free or unrestricted public access to most beaches without purchasing your little piece of the shoreline for the day.

It's just the way it is and people don't seem to mind. The result is a colorful sense of order along the shoreline. Each concession and their stretch of beach has its own distinctly colored umbrellas and lounge chairs, connecting everyone in a quaint sort of way.

While we do have a fee to drive into some of our state beaches, we are indeed fortunate to have hundreds of miles of California shoreline that we can get to without paying a dime. Because of our underlying geology and the many rivers and streams that provide large volumes of sand to nourish our beaches, we have no shortage of accessible beaches.

And while free beach access and a place to spread out your blanket or towel has always been something we take for granted in California, it's not this way along much of the Mediterranean coast. We were surprised to have to pay 15 Euros ($18 dollars) to have a place to sit for the day on a gravel beach, but with the oppressive heat all around and the cool water beckoning, we were quite happy to pay the going price.

PART V

Waves, Currents & Sea-Level Rise

Article 1

Sunlight and Sea Level
What's the Connection?

I feel compelled to first answer the question I ended my last column with, because several of you have asked - "was it ice"? Is that your final answer? It was ice.

The 10,000,000 cubic miles of sea water that were removed from the oceans 20,000 years ago, that lowered sea level 400 feet to expose the offshore continental shelf, ended up as ice sheets and glaciers that covered large parts of North America and northern Europe. Glaciers scoured Yosemite Valley. Seattle was buried under hundreds of feet of ice. Cape Cod consists of all the sand and gravel left behind as the glaciers covering New England melted and retreated, and vast glaciers scoured out the Great Lakes and extended from the northern Canada to Kansas.

We have had many periods of glaciation throughout the history of the Earth with the last interval beginning about 3 million years ago. Glacial periods have alternated with warmer intervals over time periods of about 20,000 to 40,000 years. During each of the glacial periods the water that formed all that ice came out of the oceans. So sea level has risen and fallen hundreds of feet repeatedly in concert with these periods of glaciation.

The next question is why? Why Ice Ages? Why periods of major climate change, global warming and cooling, way before we had cars and freeway and power plants to produce greenhouse gases? Stay with me here. There are several different factors that together affect whether or not we have an Ice Age and how severe it is. The most fundamental of these, the pace maker so to speak, is the delivery of heat from the sun. And it's not constant, but varies over some well-understood cycles that are determined by how far the Earth is away from the sun, and therefore how much heat we get.

The Earth's orbit around the sun isn't circular but is an ellipse, so over a period of about 100,000 years we move a little closer to the sun. We then get a little farther away, which makes things here a bit warmer or colder. The Earth also tilts slightly on its axis, which produces the seasons. But over a period of about 41,000 years this tilt changes by a few degrees, which also affects the amount of sunlight reaching different parts of the Earth. The third piece of the puzzle is a wobble in the Earth's rotation, which changes over a cycle of about 21,000 years.

These three irregularities in the Earth's orbit have been taking place throughout the 4.6 billion year history of the Earth, and in combination, determine how much heat we receive from that burning mass 93 million miles away. When combined with a few other global changes, like how the continents and oceans are distributed around the planet, where there are large mountain ranges and how high they are, the Earth's overall temperature can change by as much as 6 degrees C. This is enough to help initiate or end an Ice Age, change sea level by hundreds of feet, and make the difference between walking or sailing to the Farallons.

Article 2

Keeping Track of Sea Levels

The last 50 to 100 years of sea-level rise have been well documented by tide gages along the state's coastline from Crescent City to San Diego. The historic change in sea level at any location, however, is due to the combined effects of both how the level of the ocean has changed globally, and what the adjacent land has been doing. Nationwide (http://co-ops.nos.noaa.gov/sltrends/sltrends.shtml), you can see that, at least in terms of sea-level rise, you are much better off living in Skagway, Alaska than in Grand Isle, Louisiana.

Thousands of feet of ice depressed Alaska during the last Ice Age. As the ice gradually melted, Alaska as well as Scandinavia and other high latitude regions have slowly rebounded. This is somewhat like sitting on the edge of your mattress and then watching it rebound after you get up- only in slow motion. Because the land is rebounding faster than sea level is rising, the tide gage at Skagway shows sea level is actually dropping relative to land at 17.1 millimeters each year (nearly ¾ of an inch), equal to five and a half feet over 100 years. Needless to say, Skagway and much of Alaska aren't worrying much about sea-level rise.

In Grand Isle, Louisiana, not far from New Orleans, sea level is rising at 9.2 mm/yr, equivalent to 3 feet over a century, the highest rate in the nation. Land is sinking in the Mississippi delta area, in large part due to the deposition of thousands of feet of sediments from the Mississippi River over millions of years, which is weighing down the crust. Settlement of the peat or organic soils and ground water extraction also contribute to the subsidence problem. Ongoing sinking of the land surface combined with continuing sea-level rise is already a problem for New Orleans, and its only going to get worse. It's risky living below sea level protected by old levees, and large parts of the Sacramento delta region share this concern.

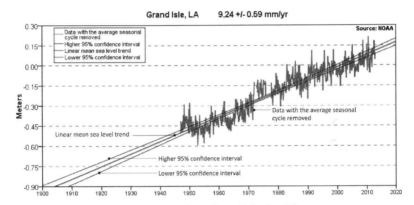

Tide Gage Record from Grand Isle, Louisiana.
(courtesy: the National Oceanic & Atmospheric Administration)

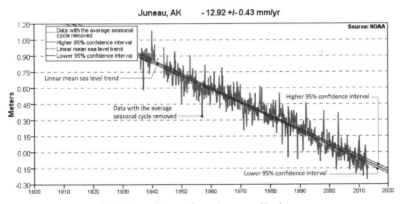

Tide Gage Record from Juneau, Alaska.
(courtesy: the National Oceanic & Atmospheric Administration)

What's been going on with sea level along the California coast? With the exception of Crescent City, near the Oregon border, and Eureka, sea-level rise rates from tide gages along the rest of the California coast are equivalent to 3 to 9 inches per century, not a very large range. Remember these values represent the combined effects of sea level increase and the motion of the adjacent land.

With a global sea-level rise of about 8 inches over the past 100 years, or about 2 mm/yr, California's coast doesn't seem to have been moving up or down much over the past century.

The landscape has been more active along the state's north coast, however, with Eureka sinking and measuring a sea level rise of 1.5 feet/100 years. Crescent City, close to the Oregon border, has actually recorded a slight drop in sea level, equivalent to about 2.5 inches per 100 years, indicating that the land is being uplifted faster than sea level has been rising.

The approximately 1.7 mm/yr rate of global sea-level rise over the last century was based on averaging out hundreds of tide gage records around the world. Beginning in 1993, we began to use satellites to measure the rate of sea level rise directly (a process known as satellite altimetry), in order to remove the land movement effect. Over the past 20 years the rate has increased to about 3.1 mm/yr, about 80% larger than the average value over the 20th century. We don't yet know for certain that this represents a long-term increase in the rate of sea level rise, but all signals are pointing in that direction.

Article 3

Return to Pleasure Point-Again

Effluent from East Cliff sewer outfall surfacing amidst the surfers, 1972.
(original photo: State Department of Boating & Waterways,
restoration: © 2004 Kenneth & Gabrielle Adelman,
California Coastal Records Project, www.californiacoastline.org)

For several months in 2010 my writing was deep into coastal erosion and seawalls, and then Sandy Lydon and I took this hike around the bay and my column went sideways for four weeks. So I'm back on track, although I'm not so sure that there is such a thing in the writing of newspaper columns as being on track.

Forty-six years ago, I arrived in Santa Cruz to join the early faculty at UC Santa Cruz. After four undergraduate years surfing at UCSB, I felt a bit landlocked in graduate school at Oregon State University. I was looking forward to being on the coast again and getting back in the water. I'd heard about Pleasure Point, so not long after arriving in town I threw my old longboard in the car and took off for the east side.

I immediately discovered several things. The first was that the water was a whole lot colder than Santa Barbara, where no one wore wet suits simply because there weren't many around in the early 1960's. We all just got out of the water about December and then started surfing again in March when the water was tolerable. But I quickly realized in my mild hypothermia that surfing here was going to require some insulation. So I bought my first wet suit from Jack O'Neill in 1968, which I still have by the way.

I also noticed that the other surfers in the water had leashes, which also hadn't been around in my Santa Barbara days. It wasn't a big deal to swim after your board when the water was reasonably warm and there weren't any rocks. But I soon realized after losing my board a few times and swimming through the ice water to get it back that the leash must have been invented in Santa Cruz. Necessity or survival was the mother of invention.

The other thing that got my attention as soon as I paddled out was the smell and discolored water right off Pleasure Point. I had minored in Civil Engineering in graduate school and had taken an interesting course called Sanitary Engineering, which is really all about sewage treatment - and the odor that day was unmistakably that of sewage. I soon discovered that in addition to being called Pleasure Point, the spot was also known by locals as Sewers.

This experience and a concern about the potential problems of nearshore discharge of treated sewage got me involved in my initial ocean research in Santa Cruz, which focused on wastewater disposal and ocean currents around Monterey Bay. In the late 1960's the wastewater from Capitola and the Live Oak area was put through primary treatment (which is basically several hours of settling and chlorination) and discharged about 200 feet off of Pleasure Point in several feet of water. In general, depending upon the nature and volume of the discharge, the level of treatment and disinfection, the distance offshore where the outfall terminates, and the circulation and mixing at that point, as well as the recreational use of the discharge area, there can be significant health risks.

After several years of study and increasing public health concerns, the discharge at Sewers was terminated and waste water from the east side was piped to the Santa Cruz Wastewater Treatment Plant in Nearys Lagoon. At that time, Santa Cruz used only primary treatment (settling and chlorination) and discharged the treated effluent 2000 feet of West Cliff Drive at the end of Almar Avenue. On a calm day you could see the cloudy discharge emerge at the ocean surface.

A hundred million dollars and several years later, the Santa Cruz plant had been expanded and upgraded, the treatment level increased to secondary level, and a new outfall pipe constructed that took the effluent several miles offshore of Natural Bridges. This investment greatly improved the situation that existed at that time and eliminated any water quality issues from both the Pleasure Point and West Cliff areas. Sewers was fortunately gone, but we still had an eroding bluff to contend with.

Article 4

Giant Waves at Sea

USS Ramapo, 1943.
(courtesy: Navy Yard Mare Island via Wikipedia Commons)

I think many of us have a fascination with extremes, which is what makes the Guinness Book of Records so interesting and bizarre. Two books have appeared recently on the topic of extreme waves: The Wave: In Pursuit of the Rogues, Freaks, and Giants of the Ocean by Susan Casey, and The Power of the Sea: Tsunamis, Storm Surges, Rogue Waves, and Our Quest to Predict Disasters, by Bruce Parker. You have to admit that those titles grab your attention, especially living in Santa Cruz, with Mavericks an hour's drive north and Ghost Tree about the same distance to the south.

Perhaps 20 years ago, most of us, even those who study the ocean, had never heard of either. But Mavericks has become a household word on the central coast, and Ghost Tree, off Pescadero Point on the 17-Mile Drive, is becoming legendary for huge surf as well. Sadly, big wave riders have died at both, and there is a group of surfers who wait patiently for either to go off each year. Personally, having spent some of my younger years surfing, they scare the heck out of me, and I have absolutely no desire to get out in waves of this size. But there is a fascination to a unique cadre of surfers with riding waves 30 or 40 feet high, or larger.

This raises several interesting questions, but I'll avoid the human desire question, and instead deal with the waves themselves. How big or high can waves actually get? Well, it depends upon what kinds of waves we're talking about, which is the subject of the two recent books.

If we are talking about surfing and waves that we typically see breaking at Mavericks or Steamer Lane, they are all generated by the wind. Wind blowing across the ocean transfers energy to the sea surface, initially forming small ripples, which over time, if the wind persists, will begin to form discrete waves. The amount of energy transmitted to the ocean surface and, ultimately the height of the waves, depends upon the velocity of the wind, the length of time the wind blows, and the fetch or distance over which the wind blows.

In order to get big waves, we need to have high velocity wind persist for an extended period of time over broad stretches of ocean. While you can get small waves in a pond or a lake, it takes a large water body, like the Pacific Ocean, to crank up really big waves. And how big can wind waves get? Its a difficult question to answer, because even most oceanographers that study waves don't have a desire to spend months at sea in modest-sized research vessels looking for the world's largest waves.

The largest wave actually observed and reliably measured to date was in the Pacific in February 1933 and happened under a unique set of circumstances. The USS Ramapo, a 478-foot Navy tanker, found itself in an extraordinary storm on its way from Manila to San Diego. The storm lasted 7 days and stretched from Asia to New York, producing strong winds over thousands of miles of unobstructed ocean.

By triangulation on the ship's superstructure as it went down the face of the wave, they measured a wave height of 112 feet from peak to trough. The wavelength or distance between two successive crests was 1,100 feet. Wave speed, which is dependent on the wavelength, was calculated at over 50 miles/hour.

The crew of the Ramapo measured the waves and lived to tell about it because their relatively short ship (478 feet) rode over these very long wavelength mountains of water without severe stress. Some of today's much longer vessels have had serious problems in seas of this size, where the length of the ship is similar to the lengths of the waves they are passing through.

The USS Ramapo made numerous runs back and forth across the Pacific and to Alaska during World War II, providing fuel to other Navy vessels. On December 27, 1942 the Ramapo was credited with rescuing the entire crew of the USS Wasmuth, a minesweeper, in heavy seas off Alaska. The Wasmuth was escorting a convey through a large Alaskan storm when two depth charges were ripped from their racks by the huge waves. They fell overboard and exploded beneath the ship's fantail. The blasts carried away part of the stern and the ship began taking on water quickly. Despite the heavy seas, the Ramapo came alongside and for three and a half hours, the tanker remained with the sinking ship and saved her entire crew. There are more big wave stories to come.

Article 5

Rogue Waves

The collapsed flight deck of the USS Bennington, 1945.
(courtesy: Australian War Memorial Collection via Wikipedia Commons)

The German container ship MS München left Bremerhaven, Germany, on a cold day in late 1978 headed for Savannah, Georgia. On December 12, the ship, two and a half football fields long and described as unsinkable, vanished with one unintelligible distress call. All that was found in a wide search of the general area was some scattered debris and an unlaunched lifeboat that was originally secured on the deck 65 feet above the water line. Its attachment pins had been "twisted as though hit by an extreme force". The best guess at the time was that the ship had been struck by a very large wave.

While seaman for many years have described huge waves or walls of water at sea, they weren't usually given much credence until recently. Encounters with such large waves have become more frequent over the past 15 years or so, however, indicating that perhaps these weren't all just sailor's exaggerations or nightmares.

In February 1995, the Queen Elizabeth II encountered what was described as a 95-foot wall of water in the North Atlantic. The ship's captain said it "came out of the darkness" and "looked like the White Cliffs of Dover". He was able to determine the wave's height because the crest was level with the ship's bridge. The wave broke over the bow with explosive force and smashed many of the windows and part of its forward deck. That same year, an oil platform in the North Sea with a wave gauge measured a single rogue wave with a height of 84 feet, for the first time.

Two cruise ships that take tourists across the South Atlantic to Antarctica, the MS Bremen and the Caledonian Star, collided with rogue waves nearly 100-feet high within a week of each other in early 2001. Both vessels had their bridge windows broken and the Bremen drifted without navigation or propulsion for two hours. The First Officer of the Caledonian Star stated that it was "just like a mountain, a wall of water coming against us".

Just four years later, in April 2005, The Norwegian Dawn, a cruise ship with 2,500 passengers on board, encountered rough seas between Miami and New York City, including three 70-foot high waves. Windows were smashed on the 9th and 10th decks, and 60 cabins were flooded, although damage otherwise was minor. In March 2007, Holland America's cruise ship, MS Prinsendam, was hit by a 70-foot high wave in the Antarctic part of a voyage around the tip of South America.

In the 21 years between 1981 and 2001, 124 ships over 600 feet long were report to have sunk, often by what is usually called "severe weather". That's a ship lost every two months!

A 5-year project was initiated by the European Space Agency in 2000 to look into how common rogue waves might actually be and if they might explain the losses of these large ships. Using twin satellites that use radar to observe waves at the sea surface, scientists initially evaluated 30,000 images for a 3-week period when the Bremen and the Caledonian Star were damaged. Even though this was a brief period of time, the team of scientists identified 10 giant waves from around the world oceans that were over 80 feet in height! This came as a big surprise and provided strong evidence that large rogue waves are far more common than was previously believed.

This initial series of observations also revealed that these giant waves often occur where ordinary wind waves encounter ocean currents. The strength of the current seems to concentrate the wave energy, much like a lens will concentrate light. These conditions normally occur far out to sea, however, so no need to worry about rogue waves attacking you on the beach.

Article 6

Lost Cargo Tracks Ocean Currents

Container ship Rena grounded off the New Zealand coast, 2011.
(photo: NZ Defence Force licensed under CC BY 2.0 via flickr)

The currents along our shoreline move things around in many different directions, sometimes north, sometimes south and sometimes onshore or offshore, depending upon the season and the winds. Whether nutrients or plankton, pollutants or lost people, oil spills or flotsam, the ocean currents can transport stuff thousands of miles. Surface currents are driven by the wind, and overall, the current patterns in the Pacific, the Atlantic and the Indian oceans are pretty similar, and reasonably well understood. Although the current patterns in the North Pacific will eventually bring any remaining Japanese tsunami debris towards the west coast, as of summer 2014, very little had arrived because the North Pacific Drift is a relatively slow current. The Japanese fishing floats found periodically along the beaches of California and Oregon are evidence of the general circulation pattern, however.

On January 29, 1992, a large container ship left Hong Kong, destined for Tacoma, Washington. Several thousand miles into the voyage, near the International Date Line, the ship hit severe storm conditions, not too uncommon in the Pacific in winter, and lost twelve big shipping containers. One of these was stuffed full of 29,000 Chinese-made plastic bath toys: yellow ducks, blue turtles, green frogs and red beavers. When the container broke open, the floating toys were carried off by the currents, and in the following years, dispersed, traveled widely, and reached some surprisingly distant places.

Their first landfall came 10 months later when over a hundred of the toys washed onto the beaches of southeast Alaska. In the fall of 1995, nearly four years later, after apparently circling counterclockwise around the entire north Pacific and passing over the original drop site, several thousand of the floaters passed through the Bering Straits between Alaska and the Soviet Union. This group continued, passing eastward into the Arctic Ocean, where they were trapped in the ice. Simultaneously, another large group of the bath toys floated south, ending up on the coastline of South America and also on beaches of Indonesia and Australia.

Moving with the ice, some of the Arctic travelers began reaching the Atlantic Ocean in 2000, where they began to thaw out and move southward. Soon plastic ducks were seen bobbing in the waves from Maine to Massachusetts. In 2001, ten years after they were released, some of them were tracked in the area where the Titanic sank 90 years earlier.

Each of the toys was labeled with the words "The First Years", the name of the Chinese company that manufactured the bath toys, so the source of any animals recovered could be confirmed. In 2003, a lawyer vacationing in the New Hebrides, off the west coast of England discovered the first faded green frog from the original spill. In 2007 a retired school teacher believed she had found the first of the toys to wash up on the coast of England, 15 years after their initial release; but under careful examination, it turned out not to have been one of the original toys. Bleached by sun and seawater, the ducks and beavers have faded to white, but the turtles and frogs have retained their original color. Two children's books have been written about the amazing voyage of the ducks and their companions, and the toys themselves have become collector's items, fetching up to $1,000 each

This wasn't the first time this has happened, however, and oceanographers were treated to some unexpected information on ocean currents. In May 1990, almost two years before "The First Years" toys started their voyage, another container ship, the Hansa Carrier, encountered a severe storm on its route from Korea to the west coast, losing 21 shipping containers. Four of the containers, holding about 60,000 Nike shoes, broke open and released the tennis shoes into the North Pacific. The following winter, individual shoes started washing up on the coast of Oregon, Washington and onto the beaches of Vancouver Island and the Queen Charlotte Islands. Some of the shoes were diverted south along the California coast, and then were carried by westward flowing equatorial currents, ending up in the Hawaiian Islands by the summer of 1992.

About 90% of all of the non-bulk cargo transported around the world is now shipped in containers. Right now, as you are reading this, there are five or six million of these containers transiting the world's oceans. And about once an hour, on average, one of those containers breaks loose and falls overboard, although usually more than one falls at a time. Some sink, some float to become navigational hazards, but they are all full of stuff. When they break open, the floating cargo has the potential to provide us with more interesting information on which way ocean currents move and how fast.

ARTICLE 7

Floating Frisbees and Coastal Currents

Drifter card.
(© 1970 Gary Griggs)

Forty-five years ago, Santa Cruz had some local water quality problems, mostly from older sewer outfalls that were located too close to the beach. In an earlier column I wrote about Sewer Peak, or the old East Cliff outfall that discharged just 200 feet off Pleasure Point. The Santa Cruz outfall was a bit longer, 2000 feet off the end of Almar Avenue, but still too close for swimmers and surfers on the West Side when the surface waters were moving towards Steamer Lane.

Both of those old outfalls were subsequently eliminated, the level of treatment improved from primary to secondary, and a new outfall line was built that takes treated waste water from both of the plants several miles offshore of Natural Bridges. Before figuring out where to put these outfalls 45 years ago, however, we needed to find out which way the coastal currents flowed so treated water was discharged in an appropriate place.

In 1970 there was also a proposal to build what would have been the nation's largest nuclear power plant on the coast just north of Davenport. Knowing that the nearshore waters would be used for the discharge of very large volumes of cooling water raised another coastal current question, which way was that effluent going to go?

Today we typically use a land-based radar system (CODAR - Coastal Ocean Dynamics Application Radar) to track ocean currents along the entire coast of California in real time, and don't even have to leave the comfort of our office and computer screen. While we now have some very sophisticated tools for monitoring which direction the coastal waters are moving, this wasn't the case in 1970. My personal interest and scientific curiosity at the time for nearshore water quality and where any present or future discharges would end up led me to undertake a study of currents around Monterey Bay. I wanted to know where all that stuff was going and where any new discharges were going to end up.

Being a bit short of funding at that early stage in my career as a young Assistant Professor, I opted for an inexpensive and simple approach. There were plastic current drifters commercially available that could be dropped in the water, and which behaved just like the notes in bottles cast adrift by shipwrecked sailors or curious children. The drifters looked like small Frisbees, each had a tail and a small brass weight that caused it to sink and move with the bottom currents. But by cutting off the weights, the Frisbees floated and moved with the surface currents where most of the action was.

We stapled preprinted waterproof postcards to each drifter, which included a catalogued number, a return address and postage, and a request to write down where and when the drifter was picked up, and then mail it back. In order to get a regional picture of Monterey Bay currents, I decided to drop the drifters monthly along a series of transect lines extending offshore from Davenport,

Almar Avenue where the Santa Cruz outfall discharged, Pleasure Point or Sewer Peak, and also Moss Landing. Six months into the study, I decided to extend the transects northward to Año Nuevo, and southward to Point Pinos and Point Cypress.

In order to make the process of dropping the drifters as efficient as possible, rather than using a small boat, which might have taken several days, I decided to go with an airplane. Another factor tipping the scale for me personally in favor of using a plane was that I had planned at least a 12-month study, and the sea conditions along the central coast are not always ideal for spending two entire days at sea in a small boat every month.

So the next step was locating a pilot and plane. I knew a graduate student who had recently flown along the coast and taken pictures, so asked him who the pilot was. He explained that the night janitor in the building occasionally rented a plane and flew out of the old Scotts Valley airport. Today doing University business is far more complicated, with requirements for commercial licenses and multi-million dollar liability policies, but this was 1970, so I called him and we set up the first aerial drop of the drifters for the next week - no approvals and no insurance. We were flying by the seat of our pants.

I bundled the surface and bottom drifters into packets of 10, held together by inserting their plastic tails into a salt ring, which would allow them to fall together after being dropped from the plane. When they hit the water, the salt would dissolve letting the weighted drifters sink to the bottom while the unweighted drifters would move with surface currents. So far, so good.

While this all seemed pretty straightforward and simple in principle, it got a little more complicated when the plane was flying over the waves at 80 or 90 miles an hour and we had to open up the door and drop the bundles at precise locations. The janitor was a nice weekend pilot, but I was asking him to do some things that I soon discovered weren't particularly safe or smart. I also was a bit surprised when we got into the plane on our first flight and he got out an instruction sheet explaining how to start up the plane. This did not instill in me a large amount of confidence in his experience as a pilot, but I later realized that this is pretty routine practice for virtually all pilots, even those who fly commercially.

Flying perhaps 50 feet above the ocean, typically under windy conditions, opening up the door repeatedly and dropping drifters out, proved on more than one occasion to be life-threatening. The incident I recall best was on our 2nd or 3rd trip, when the pilot suddenly turned around with a frightened look on his face, and asked "Can you see the tail of the plane? The aircraft isn't operating properly". These are not the words you want to hear when you are two miles off Davenport with the cold ocean just a few feet below you. I had to tell him I couldn't see the tail of the plane.

It turns out that one of the bundles of drifters I had just dropped didn't make it into the ocean, but was wedged into the elevator or that movable part of the tail that allows the airplane to move up or down. This was not a good thing. We didn't know what the problem was, however, until we had returned to Scotts Valley, minus use of the elevators, and made an emergency landing.

This neat and inexpensive research project I had dreamed up to find out which way the coastal currents were going had suddenly became much more difficult and dangerous than I had originally envisioned.

Article 8

Aerial (mis)Adventures

Studying local ocean currents 45 years ago by dropping drifters from a small plane every month for a year and a half turned out to be a regular adventure. It was important to know which way the nearshore waters moved throughout the year, although the value of the information wouldn't become clear for some years to come. I'll get to that in my next column, but as a young scientist trying to start a coastal research project, there were more mishaps ahead. The process of doing research is often more interesting and exciting than the results.

After surviving the first emergency, the janitor turned pilot said he had a solution to avoid getting the bundles of current drifters caught on the tail of the plane. An experienced pilot can adjust the controls of the plane so that it flies in a skid, called a crab. This is one way of landing a plane in a strong crosswind.

The pilot was proposing to fly the plane in a skid, with the tail off to one side so the drifters didn't get stuck on the tail. Not knowing much about flying, I said sure. Let's go for it. Had I known that this wasn't particularly safe, I might have said something else, but I had another drop to make in two weeks.

The approach I worked out for dropping the drifters was to fly at a known speed, usually about 80-90 miles per hour, and then use a stop watch to determine when we were ½ mile, 1 mile, 1 ½ miles, 2 miles and 3 miles offshore, at our established offshore drop points.

To manage the drop each month required an assistant, who sat in the back of the plane and handed me the pre-marked bundles of drifters. I would then open the door and drop these at the appropriate times to hit our targets. I had a younger brother who was a UCSC student at the time, and he was usually willing to go along. But, depending upon the day and time, I grabbed whomever I could find who wanted to go flying. I didn't bother to explain to them that there were life-threatening risks, that we didn't have any insurance, or that we had already experienced some serious mishaps. I needed to get the next batch of drifters in the water.

We actually did drop the drifters while skidding the plane for a while without any major incidents, until about our sixth month. Back in the early 1970's, Scotts Valley had a small airport, right alongside Mt. Hermon Road directly in front of what is now the King's Village Shopping Center. There were two hazards for pilots to contend with at Scotts Valley, one when landing and the other when taking off. Next to the landing strip was a propane gas company lot with lots of large tanks full of flammable gas. Having this adjacent to the runway always seemed like a bad combination. Taking off to the west, there was a ridge directly in the flight path so you had to climb quickly to avoid becoming a statistic.

About month six I grabbed a student who had been working for me getting the drifters ready to launch each month. He was excited about flying so agreed to come along. He was a very large guy, probably weighing 250 pounds or so. The pilot was also a large man, although I didn't think too much about their combined weight until we left the ground.

In less than 15 seconds a flashing red light came on, which was the stall light indicating the plane was climbing too quickly for the load. With a mountain side directly in front of us and a flashing red light in my face, I again had that fatal feeling that we were in deep trouble and that maybe I should have used the boat after all. Being seasick for a two days every month all of a sudden sounded better than dying in a plane crash.

My janitor pilot was clearly a little alarmed, but by banking right and heading up Lockhart Gulch, he was able to avoid colliding with the low ridge in front of us. He reduced the rate of climb enough to keep the stall light off, and we slowly gained enough elevation to just clear the ridge top without hitting any trees.

The rest of that flight went off more or less as planned although I was still uneasy about using the crab or skid approach. I talked to my brother about my concerns and he told me his neighbor was a commercial airline pilot and might be worth talking to. I met Rich the next weekend and explained the coastal current study, our several near disasters, and then the crab solution. I recall his eyebrows going up, and with astonishment saying – "You're doing what?"

He explained that this was not a safe maneuver under any conditions, particularly with the plane door open, flying at low elevations directly above the water with strong winds. He had his own plane, which had a V-shaped tail that would avoid the drifter release problems, and also had more power than the rental plane we were using. He was also a professional pilot with thousands of flying hours. My research project was starting to sound safer.

The only real challenge was that his plane had only one door, and that was on the pilot's side. But I couldn't very well ask him to open his door and drop the drifters while he was also trying to fly the plane. The window on the passenger side was hinged but only opened a few inches. But Rich and I were able to take a pin out of the latch system so it could swing freely upward and I could get the drifters out. Opening the window at 80 or 90 mph meant that there was a lot of lift on the window, however, so we attached a leash to hold it down between drops. This was important.

The stage was set for our next near disaster. The assistant in the back seat had to hold the leash tightly to keep control of the window while I was using both hands to drop the drifters. On one of our first flights, my helper on the flight accidentally let go of the leash, the window swung loose and began hammering the top of the plane. This not only made a loud racket, but was also of considerable concern to the pilot. I remember him yelling- "WHAT'S THAT?"… followed by "GRAB THE WINDOW!" So three miles off Moss Landing, I had to reach the upper half of my body out of the plane and grab the loose window, which was still slamming against the roof of the plane, and hold it down until he could make an emergency landing in Monterey. Hanging out the window for the next 15 minutes I had plenty of time to again wonder if there might be a safer way to find out which way our coastal currents were going.

Article 9

Coastal Currents and Missing Sailors

Brothers Bill & Matt Hopps (center) and their rescuers
Don Kinnamon and Cary Smith of the Santa Cruz Harbor staff.
(courtesy: Hopps family)

We dropped drifters into the ocean monthly around Monterey Bay from November 1971 to April 1973 in hopes of finding out which way the coastal currents moved. Most of the flights were without incident once we resolved how to fly the plane and drop objects out of it safely.

What we did discover was that nearshore currents patterns along our coast change seasonally. The winter months are characterized by a well-defined northerly flow known as the Davidson Current, which moves at speeds of about two to six miles per day. Drifters dropped between November and March typically headed north, and were usually recovered along the beaches from Davenport to Pacifica. The greatest distance traveled by one of the little red Frisbees was 750 miles, ending up north of the Columbia River at Ocean Shores, Washington. This doesn't compare with the little yellow ducks that came out of the shipping containers, but then we only recovered about 30% of all the drifters we dropped, so I don't actually know how far the others may have eventually traveled.

An abrupt reversal occurs in the spring as nearly all drifters dropped in those months moved southward with the California Current. From March to May 1972, 93% of the surface drifters moved southward or onshore. Those dropped off of Santa Cruz and Pleasure Point usually ended up along the beaches of Monterey Bay, while those dropped offshore of the Monterey Peninsula often traveled north and also ended up in the bay. Typical California Current speeds were also about two to six miles per day, although speeds as high as 12 miles a day were also recorded. The longest southerly voyage was nearly to Morro Bay, a 100-mile trip from the Monterey Peninsula drop point.

Nearshore circulation in summer and fall was less regular, with about equal number of drifters moving north and south, which seemed to be due to eddies or gyres within the larger California Current system. The late spring and early summer months are part of the upwelling period and usually characterized by winds from the northwest, and due to the Earth's rotation, surface waters tend to move offshore. Nutrient rich bottom water rises to replace the surface waters in a process known as upwelling. This is common along the eastern sides of all oceans and produces very high biological productivity. The nutrients fertilize diatom and algal blooms, followed by krill and other small planktonic organisms, which are in turn fed on by sardines and anchovies, and then the larger fish, marine mammals and birds. Our current studies were helping us to better understand how things actually moved around offshore, but I didn't realize then how the information we collected might ultimately be used.

On Labor Day 2002, two distraught young ladies approached the Santa Cruz Harbor office nearly at dark. They were very concerned that two brothers, the husband of one lady and the fiancé of the other, had taken out a Hobie Cat four hours earlier on a very windy afternoon and hadn't yet returned. The Coast Guard was notified, took charge and along with the Harbormaster staff, launched a full-scale search and rescue operation. The wind at the time was from the northwest so all the rescue teams and boats went southeast toward Capitola and Rio Del Mar. By midnight, despite their best efforts, they had no contact and decided to give up the search and resume it again at daylight.

Don Kinnamon, who was on the Harbor Master staff and part of the search crew, returned to the office very discouraged. But remembering the current study I had done 30 years earlier, he recalled that currents were usually more irregular in the early fall and could be moving in a different direction than the wind was blowing, even offshore. He called another crewmember, Cary Smith, now a deputy harbormaster at Pillar Point Marina, who had just gotten home after a long shift. They decided to launch a second effort at midnight. After informing the Coast Guard and the Harbor Master, they decided to head offshore 210 to 220 degrees, or southwest rather than southeast, where the earlier search had been focused.

About ten miles offshore, on a pitch-black night, in a slow series of sweeps, Don and Cary heard a voice calling out. They were soon able to locate one of the brothers, Bill Hopps, whose core body temperature was very low and whom they believed was within minutes of dying. His first weak words were "find my brother". Don and Cary struggled with the decision of trying to warm him up and keep looking for the other brother, or race back 10 miles to the harbor to get him to an ambulance and medical attention.

They decided to make the run to shore to drop off Bill. After leaving him with a medical team, they turned around and retraced their route over the now pitch black ocean to their previous location. They continued their slow sweeps back and forth, and about a mile farther offshore found the second brother, Matt, just barely alive.

Several weeks later, the second brother to be rescued that night was getting married to his fiancé in Los Gatos. They invited Don and Cary to the wedding. Instead of toasting his new wife, Matt explained to the wedding guests that he wanted to instead make the first toast to the two men standing in the back, who saved his life and made the wedding possible.

While at the wedding that afternoon in 2002, Don Kinnamon met one of the bridesmaids. They subsequently became partners in life and now have a young son. While this wasn't an outcome I could ever have predicted, it made the coastal current project all the more interesting and rewarding.

Article 10

Sea-Level Rise-Should We Worry?

Sea level rise probably isn't an issue on most people's list of top ten worries. It's usually the catastrophic or instantaneous rise in water level such as occurred in Japan during the devastating tsunami of March 11, 2011 that we notice. It certainly got our attention at the Santa Cruz harbor the next day, at least for a while.

There is also a much slower, glacial-paced rise in sea level that is beginning to be of concern to those who need to worry about the elevation of the ocean surface, which also determines the location of the shoreline. One good example is the operators of the San Francisco International Airport, well and also the pilots that land and take off from SFO every few minutes.

The airport originally opened in 1927 as Mills Field Municipal Airport on 150 acres of cow pasture leased from owner Ogden Mills. It was renamed San Francisco Municipal Airport in 1931 and Municipal was replaced by International in 1955.

The cow pasture was originally tidelands and over the years the airport was repeatedly expanded as fill was placed out into the bay. As a result, today the runways are just a little more than a foot above the highest tides.

Much of the shallow, marshland around the margins of San Francisco Bay was filled over the past century for a variety of uses that made sense at the time, including housing developments such as Foster City and Redwood Shores (with no redwoods), freeways, stadiums and airports. Because transporting and placing engineered fill is expensive, there was no reason at the time to add more fill than was necessary to get the ground level just above the highest tides.

Part V: Waves, Currents & Sea-Level Rise

San Francisco's Bay Conservation and Development Commission (BCDC), a governmental agency that plans for and has permit authority for how land is used around the shoreline of San Francisco Bay, was one of the first agencies in California to be concerned about rising sea levels. In 2005 they carried out a study to determine how a sea-level rise of three feet would affect the development around the bay. Based on information available at that time, three feet of sea-level rise was a reasonable projection for the year 2100.

The results of this analysis came as a surprise to most people. Using topographic maps with precise elevation control, it's pretty straightforward to determine areas around the bay that would be inundated by a three-foot rise in water level. This is often called the bathtub approach.

BCDC completed a second study in 2008 and included areas that would be submerged by both a 16-inch rise in sea level, which is on the higher range of recent projections for mid-century or 2050, and also 55 inches, believed to be on the higher side of the range of projections of sea level for 2100. Maps of the San Francisco Bay shoreline and areas that would be inundated with these water level increases are available on their website at: http://www.bcdc.ca.gov/planning/climate_change/index_map.shtml.

A rise in sea level of only 16-inches puts the SFO runways underwater.

You may be saying to yourself, no worries, I'll just use Oakland International. Well, the Oakland airport, which coincidentally also opened in 1927, was built on former tidelands as well. It has a long and interesting history, including being the departing point for Amelia Earhart's final flight in 1937. But Oakland's runways are also submerged with a 16-inch rise in sea level.

So when do we start to get concerned or should we be doing anything? Or perhaps more appropriately, should the state of California be doing anything?

If residents along the Mississippi River know floodwaters are rising and all indications are that they are going to continue to rise, they don't wait to see how high the water is going to get before they start stacking sandbags or taking other precautions.

In November 2008, Governor Arnold Schwarzenegger signed an Executive Order focused on the impacts of climate change on the state, which included a long list of Whereas' (which for some reason is typical in such documents) followed by eight directives. Seven of the eight directives dealt directly with the issue of future sea-level rise. The first of these ordered the California Resources Agency (now led by John Laird of Santa Cruz) in cooperation with the Department of Water Resources, the California Energy Commission, the Ocean Protection Council and California's other coastal management agencies, to request the National Academy of Sciences to convene an independent sea-level rise science and policy committee made up of state, national and international experts to complete the first California Sea-Level Rise Assessment Report.

The governors of Oregon and Washington along with the U.S. Geological Survey, the Army Corps of Engineers and NOAA also signed onto this request as sponsors of the study. After requesting nominations for the committee members from a wide variety of scientists, public officials, agency staff and others, a long list of nominees was reviewed and vetted by the National Academy of Sciences, and a committee of 13 individuals was selected. The committee included physical oceanographers, coastal geologists and geophysicists, glaciologists and meteorologists, climatologists and coastal engineers. They represented universities and federal agencies in ten different states.

The National Academy of Sciences was established by an Act of Congress and signed by President Abraham Lincoln in 1863, at the height of the Civil War. It calls upon the National Academy to investigate, examine, experiment, and report upon any subject of science, whenever called upon to do so by any department of the government.

Since 1863, the nation's leaders have turned to the National Academy for advice on scientific and technological issues that frequently pervade policy decisions. Most of the science policy and technical work is conducted by the National Research Council (NRC), which was created specifically for this purpose and which provides a public service by working outside the framework of government and politics to ensure independent advice on issues of science, technology and medicine.

The NRC enlists committees of the nation's top scientists, engineers and other experts, all of whom volunteer their time to study specific concerns, issues or problems. The results of their deliberations have inspired some of the nation's most significant and lasting efforts to improve the health, safety, education and welfare of the nation's people.

There is a view long-held by some that science will lose its integrity, and scientists will lose their impartial reputation, if scientists stray into the domain of public policy. This can be quite comforting to some scientists because it limits their personal responsibility to their technical field of expertise. It can also be comfortable to those in the policy arena because science often presents evidence that can be perceived as a threat to vested interests.

The committees set up by the National Research Council to look into critical and timely issues are excellent examples of how science can contribute in an impartial and objective way to helping to resolve many of the pressing societal issues we face as a state and nation today. The NRC West Coast Sea-Level Rise Committee held their first meeting in January of 2011 at the Seymour Marine Discovery Center at Long Marine Laboratory.

Article 11

Sea-Level Rise-Sorting Out the Pieces

California has often been a leader in innovation and in also in stepping forward to deal with environmental issues. We took the lead in asking the National Academy of Sciences to conduct a study on the future of sea-level rise and its effects along the California coast. Governors of Oregon and Washington then joined in sponsoring the study.

Much of the initial concern about how future sea-level rise would affect the coast of California came from state agencies, such as Cal Trans, Parks and Recreation, State Lands Commission, Boating and Waterways, the Coastal Commission and Coastal Conservancy, and others who were concerned with how their responsibilities, their planning, their facilities or infrastructure would be affected by a rise in sea level of one, two, three or more feet over the next 50 to 100 years.

Beach level park facilities like Seacliff State Beach have regularly been battered and damaged by high tides and storm waves and subsequently rebuilt. Cal Trans has highways and bridges located within a few feet of sea level that are sometimes undermined or inundated. The Coastal Commission makes decisions monthly on whether new development should be approved along low-lying sections of shoreline or along eroding coastal bluffs, locations where retreat is an ongoing process. Pleasure Point, Opal Cliffs and Depot Hill are all local examples of eroding cliff top areas. The Venetian Court in Capitola, Pot Belly Beach, Las Olas Drive, Beach Drive, and Via Gaviota at Aptos Seascape are beach level developments, which have all been damaged by past storm events occurring at times of elevated sea levels.

As sea level continues to slowly rise, the damaging events of the sort we have witnessed in the past when large waves arrived at times of elevated water levels will occur more frequently. Property and structural damage can be expected to increase as well.

The questions that California, Oregon and Washington wanted answered by the National Research Council Committee were:
1. How much will each of the major contributors to sea-level rise (ocean warming, melting of glaciers and ice sheets, etc.) add by 2030, 2050 and 2100, and what are the uncertainties involved with each?
2. How important are regional and local contributions to sea-level rise (such as El Niño events, storms, and coastal land motion (uplift or subsidence), for example)?
3. What is known about climate related increases in the size and frequency of storms and how they might affect sea level?
4. How will different coastal environments respond to sea-level rise and increased storminess?
5. What role to different coastal environments or habitats, whether natural or restored, play in providing protection from future inundation and wave attack?

In all honesty, there have been entire books written about sea level and how and why it has changed over geologic and recent time, and it's challenging in a series of short columns to do justice to what is a messy and complicated topic. But what the heck, I might as well try, although it may take more than one column; it's a big ocean and sea level is a complex topic.

There are a handful of geologic or oceanographic processes that affect sea level at any particular location on the planet. For any of you who have been following these columns for a while, you might recall, or might not, that I wrote about some of these processes a little over five years ago in the spring of 2009. I won't repeat that discussion, but to keep it short, there are essentially two different sorts of processes that determine what sea level is doing here in Santa Cruz, or anywhere else along the West Coast for that matter.

One set of these is oceanographic, basically all the processes that affect or change the total volume of water in the oceans. These include how much of the Earth's water is tied up as ice, in Antarctica, Greenland, and all the rest of the mountain glaciers around the world. Another important consideration is the overall temperature of the oceans. Warmer water is less dense and therefore takes up more volume. A warming ocean will lead to a rising sea level.

These are long-term processes, taking place over tens to thousands of years. There are also short-term processes, the biggest ones take place twice every day, are very predictable, and can change sea level in Monterey Bay by as much as 8-9 feet over 24 hours. These are the tides. There are also strong onshore winds, atmospheric pressure differences, El Niño pulses of warm water, and the set-up created by a group of very large waves washing up on the shoreline. These short-term effects can last minutes to days.

Complicating this a bit more is vertical land motion. The Earth is alive, inside and out. Glaciers, rivers, waves and wind all move stuff around on the Earth's surface. But the interior of the Earth is moving around too, and responds to loads placed on its outer crust. The two best understood examples are the piling up of great thicknesses of ice on the continents, and the loading of tens of thousands of feet of sediment on the sea floor.

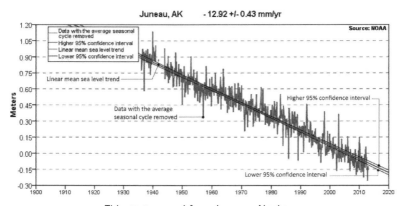

Tide gage record from Juneau, Alaska.
(courtesy: National Oceanic & Atmosphere Administration)

Part V: Waves, Currents & Sea-Level Rise

From Seattle to Cape Cod and north to the Arctic Circle, thousands of feet of ice covered much of North America during the last Ice Age. That ice weighs a lot and it depressed the land surface. As the ice melted over the past 20,000 years, not only did the addition of all that melt water raise sea level about 350-400 feet, but the northern latitude land masses have been slowly rebounding ever since, much like your mattress does when you get out of bed in the morning, only at a much slower rate.

The coastal land of Louisiana has been sinking, in part due to the deposition of thousands of feet of sediment from the Mississippi River over millions of years. All of this sand and mud built a huge delta, which gradually depressed the seafloor and the adjacent coastline as well. Additional subsidence has resulted from the compaction of organic rich sediments and also from ground water and petroleum withdrawals. This is why much of New Orleans is actually below sea level and one reason why Katrina and other hurricanes are so damaging.

Juneau, Alaska and Grand Isle, Louisiana (about 50 miles south of New Orleans) are experiencing very different vertical land motions. While sea level rose on average about 8 inches globally over the past century, it has risen at a rate of about 36 inches per century at Grand Isle because the land is sinking so fast. In

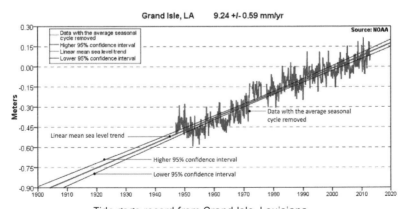

Tide gage record from Grand Isle, Louisiana.
(courtesy : National Oceanic & Atmosphere Administration)

Juneau, the tide gage shows sea level actually dropped by 50 inches over the last century because the land is still rebounding from the removal of the thick ice cover of the last Ice Age. So residents of Juneau are not much worried about a rising sea, but residents of coastal Louisiana probably have this in their top ten concerns.

California is tectonically active with plates colliding north of Cape Mendocino and moving alongside one another along the San Andreas Fault from Cape Mendocino to the Gulf of California. These processes affect whether the coastline is rising or sinking and therefore how local sea level might be changing.

So the assignment to the 13 members of the National Research Council Committee was to sort all of this out and tell us just what's likely to happen by 2030, 2050 and 2100. No problem. We'll get right on it.

Article 12

Predicting Future Sea Level

Coastal flooding at Corcoran Lagoon, Santa Cruz County, 1983.
(© 1983 Gary Griggs)

The West Coast Sea-Level Rise Committee of the National Research Council held four meetings over six months in 2011 in exotic places like Santa Cruz, Portland, Seattle and Irvine. The final report was released in early 2012 and is available free on-line at: http://www.nap.edu/catalog.php?record_id=13389

There may be some who might ask- why bother or worry about future sea-level rise? What difference will it make? I guess in some ways it's like asking any question about the future. How much should we save for our children's college expenses, or how much money will I need when I retire? Its been said that prediction is very difficult, especially about the future.

For virtually the entire 8,000 years of human civilization, sea level has been nearly constant and wasn't an issue that early humans had to deal with. This has changed for two reasons: 1) sea-level is rising and at an increasing rate; and 2) we have built so much of our present civilization in coastal regions or along shorelines, and therefore, very close to sea level. As a result, the developed world is now beginning to face problems that we never before had to deal with.

The San Francisco International Airport is just one example, but it's a big one. It's not a question of whether sea-level rise will happen; that's been going on for some time. But how fast it will happen, how high it will get, and how will this affect the shoreline. This is what the National Research Council Sea-Level Rise Committee was charged with trying to figure out. As a member of that Committee, I can report that all of the committee members worried a lot about these questions, knowing the answers had big impacts on communities along the coasts of California, Oregon and Washington, and their future planning.

There are several ways to approach these questions, with some similarity to the question of how much to save for your future retirement. You could take the simple approach and look at your overall monthly cost of living now, how it has increased over time, and extrapolate that into the future and hope you are not surprised. Or you could dissect your monthly expenses - how much you spend for housing, food, transportation, medical expenses, and all the other things that consume your paycheck - and then try to carefully evaluate each of these and how they might individually change in the future, 10, 20 or 30 years from now, and add them all up.

There are some uncertainties with each. How far out into the future do you feel comfortable simply extrapolating or extending your total monthly expenses based on the past 15 or 20 years? Or if you choose the item by item approach, how certain are you about future housing, medical expenses, food, or any of the other costs of living? What if the economy tanks and your savings or IRA disappear, or your house value drops, or social security is reduced? There are lots of unknowns.

We faced some of the same challenges in projecting sea level very far into the future. The committee took several different approaches, and used the best science available at the time of our study in an effort to develop the most reasonable and scientifically sound conclusions.

One approach is to take the rate at which sea level has risen over the past 100 years or so, and then use our best judgment to extrapolate this curve into the future. Like the funds needed for your retirement, estimates for the next 10 years aren't too much of a stretch, but at 2030, 2050 or 2100, the uncertainties get progressively larger, simply because there are so many factors affecting future sea-level rise that we can't accurately predict. Things like, how much more fossil fuel will the world's largest carbon dioxide emitters (China, USA, the European Union and India) burn before they transition to renewable energy sources. The more carbon dioxide that enters the atmosphere, the more the Earth and the oceans heat up and seawater expands, the more additional ice that will melt, and the higher sea level will rise.

Another approach is to look carefully at all the factors that contribute to sea-level rise and then try to determine how much each source has contributed historically and how these might change in the future. How fast will the ice sheets and glaciers of Antarctica and Greenland and the other smaller mountain glaciers around the world retreat or melt by some future date, and how much will each of these contribute to future sea level rise? How much warmer might the oceans get, and therefore, how much will they expand? Again, these are difficult questions to answer because there are so many uncertainties about the future of human behavior globally, as well as uncertainties about how each of these natural systems actually respond to a warming Earth.

Again, there are parallels to financial planning or savings for your retirement. There are past trends in inflation, you know what your needs and expenses are today, but you don't have a crystal ball that can predict all future costs. But because you don't have all the answers doesn't mean you don't plan as carefully as you can.

All evidence indicates that sea level will continue to rise for at least the next several hundred years, and probably longer. Tide gage records from around the world over the past century (http://tidesandcurrents.noaa.gov/sltrends/sltrends.shtml) indicate that sea level rose at an average rate of about 1.7 mm/yr, or nearly 7 inches over the last 100 years, not too much really. Satellites have provided a more precise set of observations over the past 20 years and indicate that the global rate of rise has increased to about 3.2 mm/yr, equivalent to a little over 12 inches/century. While this rate is still quite low, it represents a significant increase over what we experienced over the last century.

What does seem clear is that over at least the next 25 or 30 years that it will be the short-term increases in sea level, during large El Niño events or hurricanes, for example, combined with large storm waves and high tides, that will inflict the most damage on the coastline. The El Niño winters of 1978, 1982-83 and 1997-98 caused hundreds of millions of dollars in damage along the California coast as a result of large waves, arriving at times of high tides, when sea level was already elevated due to an El Niño driven bulge of warm water along our shoreline.

Over time, the continued rise in sea level will begin to threaten and inundate low-lying coastal areas, such as the San Francisco and Oakland International Airports. There are many areas along low-lying sections of California's coast that are flooded during extreme high tides today, and any additional rise in sea level will only exacerbate these problems.

A new website has been established (http://california.kingtides.net/) where photographs taken along the shoreline at extreme high tides (known as King Tides), which occur several times a year, can be posted. These provide some surprising images of very low-lying areas that most of us don't see under these conditions.

Eventually sea level rise will also begin to threaten many of the state's heavily used beaches. For natural or undeveloped beaches, such as Natural Bridges, or those along the interior of Monterey Bay, a rise in sea level will gradually move the beach and shoreline east or inland, as it has done for the past 20,000 years of sea-level rise. We will still have a beach, it will just move a ways inland.

Where the inner edge of the beach is fixed by a seawall, road, or buildings, however, which is the case for Cowell or Main Beach in Santa Cruz, the shoreline cannot migrate inland so a rise in sea level of 2 or 3 feet will permanently inundate the beach, as it does during large winter storms now. While this won't happen immediately, all evidence suggests that this will begin to be an increasing concern within the next 25 or 30 years. I hope I'm wrong.

Capitola Esplanade flooding, 1983.
(© 1983 Gary Griggs)

Article 13

Big Wave Surfing

Guinness World Records confirmed in May 2012 that Garrett McNamara had indeed ridden the largest wave ever surfed. In November of 2011, the 44-year old Hawaiian surfer rode down the face of a 78-foot high wave off the coast of Nazare, Portugal. He said he originally didn't want to attempt any more waves that day after wiping out a number of times on even bigger waves in the same location. For risking his life, he was awarded $15,000 for the biggest wave ridden last year. Fortunately for Garrett, he survived to collect the reward.

It took a while for the judges to officially agree on the height of the wave, which was accomplished by carefully looking at high-resolution digital images from different angles. They used McNamara's height in a crouch and the length of his shinbone and compared these to the top and bottom of the wave. Documenting the height of the largest wave every ridden has gotten pretty scientific. To claim the largest wave ever ridden, it's not enough to say, well I think it was about 75 feet high.

Huge waves fascinate many of us, and surfing big waves has evolved into one of those extreme adventure sports for a handful of surfers. Mavericks has become legendary along the central coast for huge winter waves. Ghost Tree on the 17-Mile Drive, and Nelscott Reef along the central Oregon coast are developing similar reputations for having very large waves under the right set of conditions.

Other than the occasional tsunami and the semi-daily tides, which are waves of a different sort, all the waves we see breaking along our coast, whether at Pleasure Point, Steamer Lane, or

Mavericks, are generated by the wind. Wind blowing across the ocean transfers energy to the sea surface, initially forming small ripples, which over time, if the wind persists, will grow to form distinct waves. Ultimately, the size of the waves is a function of the amount of energy transmitted to the ocean surface, which in turn is determined by the velocity of the wind, the length of time the wind blows, and the distance over which the wind blows.

In order to get really big waves, we need high velocity wind (40 to 50 mph) persisting for an extended period of time (36-48 hours) over vast stretches of ocean (500 to 1,000 miles). While you can get very small waves by blowing into your coffee cup, modest size waves in a swimming pool or lake, it takes an ocean, like the Pacific, to crank up really large waves.

We get a lot of large storms in the North Pacific, which send big waves towards the coast of California. But obviously, those waves are much larger at some places than others. The same storm south of the Aleutians will produce very different size waves at Mavericks than at Cowell. Why?

The bottom topography is the other big factor affecting how high those waves can get as they approach the coast. As waves enter shallower water, their underwater portions begin to feel the drag of the seafloor and they start to slow down. The water beneath the waves is also being compressed into shallower and shallower water, so the waves will increase in height as they get closer to shore.

Where bedrock on the seafloor, a reef, or an otherwise shallow area of seafloor is encountered, the portion of the wave passing over the high area will slow down, and the rest of the wave front will refract or bend around the rock outcrop or reef. Waves will thus wrap around points or high areas on the seafloor, concentrating energy and producing higher waves at these locations. Many of California's legendary surfing spots are located at points where wave refraction or bending concentrates wave energy so they break progressively, forming a line up. Malibu, Rincon, Pleasure Point and Lighthouse Point, to name a few, are good examples.

We don't normally use the height of a crouched surfer, or the length of their shinbone, to accurately measure wave heights. Waves have been measured accurately with offshore buoys or seafloor pressure gages for about 35 years along California's coast. There are presently 25 active gages or buoys, most of them offshore southern California. These instruments are maintained and their data recorded by the Coastal Data Information Project (CDIP) and by NOAA through their National Buoy Data Center (NBDC), which both have websites with a lot of information available on-line (http://cdip.ucsd.edu/ or http://www.ndbc.noaa.gov/)

The closest wave-recorder to Monterey Bay is a NOAA buoy, about 27 miles offshore in nearly 7000 feet of water. It has wave records extending back to 1987. These buoys are paid for by your taxes, and the information is there for your use.

If you are curious, you can dig out all sorts of interesting wave data. In the ten years between 2001 and 2011, for example, there were 99 offshore events when waves exceeded 23 feet in height for at least 3 hours. There were 31 periods when waves were at least 26 feet high for 3 hours, and on four occasions in 2008 waves were over 29 feet high. This is not Garrett McNamara's 78 foot wave, but we do get some very large waves offshore. Is the wave climate changing along the west coast and are waves getting larger?

Article 14
Are Waves Getting Bigger?

Waves overtopping Beach Drive seawall, Rio Del Mar, 1983.
(© 1983 Gary Griggs)

Big waves are one of the things that many people associate with the central coast. Mavericks and Steamer Lane are known around the world for legendary waves that attract surfers and surfing contests year after year. Surfers may be happy to know that there is some evidence that waves may be getting larger, although very slowly. Increases in wave heights from the most intense storms have occurred in both the North Atlantic and the Northeast Pacific in recent decades.

In the Atlantic this increase has been documented by the Seven Stones lightship anchored on a reef offshore from the southwest coast of England. Similar increases have been detected in the Northeast Pacific in the records from a series of 24 NOAA buoys along the west coast.

Although the information from offshore wave buoys only extends back about 30 to 35 years, there are observable increases in wave heights that are taking place. While it is not yet completely certain if these increased wave heights are related to the warming of oceans and increases in winds, there is a high probability that these are in fact related.

The causes for increased wave heights are not agreed upon yet, but might include changes in storm tracks, higher wind speeds, more intense winter storms, or other factors. Whatever the cause, the increases are important in their impact on ship and boating safety, shoreline hazards including coastal erosion, and in the engineering design of ocean and coastal structures.

The Pacific Northwest is well known for the severity of its wave climate. The largest waves off the coast of Washington were measured at the Grays Harbor buoy, north of the mouth of the Columbia River, on March 3, 1999. Offshore wave heights exceeded 29 feet for over five hours, and the highest waves recorded during that storm were nearly 35 feet. These are not waves for the faint hearted to be caught in.

Wave height increases along the west coast are greater off the northern California, Oregon and Washington coasts than offshore central and southern California. Over the past 35 years, the average wave height (designated as the significant wave height, or the average of the highest 1/3 of the waves) off Oregon has increased 1.7 feet. But, the average of the five highest waves recorded in this same area each year has increased by 8 feet during this same period. The average waves are getting larger and the biggest waves are getting larger faster.

Along the California coast, average wave heights during the 20-year period from 1980 to 2002, increased about 1.5 feet along the central California coast. Averages don't tell the whole story, however. If you have one foot in 140-degree scalding water and one foot in ice water, on average you are a reasonable 70 degrees, but you're far from comfortable.

Looking more carefully at changing wave heights along the southern California coast, from 1984 to 1995 there were seven storms that produced wave heights of 16 feet or greater and four storms that generated waves 20 feet or higher. Over the next 15 year interval from 1996 to 2010, there were 69 events with waves of 16 feet or greater and 10 events that resulted in wave heights of 20 feet or greater. Something is changing

Whether this trend of increasing wave heights will continue into the future isn't clear yet. The Monterey Bay coastline has experienced coastal storm damage at times of high tides, large storm waves, and elevated sea levels throughout at least the past 140 years of newspaper history. East Cliff, Capitola, Seacliff and Rio del Mar have all been repeatedly hammered.

When increasingly large waves are combined with a gradually rising sea level, we can expect more frequent flooding and inundation of low-lying areas and an increased rate of coastal cliff and bluff retreat. Although wave characteristics may change in the future in unknown ways, their patterns over the past several decades have been well documented. The effects of storm waves on the coast are not new to Monterey Bay or California, but these effects are likely to be gradually exacerbated by rising sea levels.

Article 15

Sea-Level Rise
Searching for the Answers

In one of his last formal acts as Governor, Arnold Schwarzenegger signed Executive Order S-13-08, which directed state agencies to plan for sea-level rise and its coastal impacts. The Order also requested the National Research Council (NRC) to establish a committee to assess future sea-level rise to inform these state efforts. The governors of Oregon and Washington, as well as 10 state and federal agencies, subsequently joined Governor Schwarzenegger in this request.

Why should the governors of California, Oregon and Washington care about a few inches of sea level rise? Don't they have more important things to worry about?

Well, the governors do have some other obvious problems, but California also has a lot of investment-billions of dollars worth, in fact-within a few feet of sea level: San Francisco and Oakland airports for starters. The runways start to go underwater with about 16 inches of sea-level rise at high tide. The next time you take off from SFO, take a look out the window and see how close the surface of the bay is to the level of the runway. It's pretty darn close.

Californians and recent transplants tend to prefer coastal counties. They historically grow the fastest, are the most densely populated, attract the most visitors, have the most jobs, and contribute the greatest amount of money to our state's economy.

There a number of state agencies that have investment and infrastructure that we all depend upon within a few feet of sea level. CalTrans, State Parks, the California Energy Commission, State Water

Part V: Waves, Currents & Sea-Level Rise

Resources Control Board, the Coastal Commission and Coastal Conservancy all need estimates and projections of sea-level rise to assess future coastal hazards and risks. Not only do they need to worry about risks to existing infrastructure but they also need to make decisions on new agency investments in transportation, energy, water, and sewage treatment infrastructure so that it won't be flooded, inundated or eroded any time soon.

State agencies as well as local governments also have to figure out how to modify their design and construction standards, and develop strategies for how they can either protect, relocate or adapt parks and parking lots, highways and bridges, railroads, power plants, sewage treatment plants, sewer lines and pump stations against coastal erosion and inundation.

The difference between whether sea level rises 6 inches by 2050 or 18 inches isn't anything to lose sleep over if you are living comfortably in the Santa Cruz Mountains 500 feet above sea level. There might be some other things to worry about in the mountains, however, like winter landslides and mudflows, or Highway 9 being closed by falling trees. But, if you happen to live in the Venetian Courts on the beach in Capitola, or along Beach Drive in Rio Del Mar, you probably think a bit more about the possibility of high tides, large waves and El Niño events knocking loudly on your ocean view windows and sliding glass doors.

In contrast to the recent law passed by the North Carolina legislature to limit sea-level rise (I'm not sure just how they plan to do that), California took a different approach nearly two years ago by asking for a blue ribbon national committee to look at what might lie ahead. As I mentioned in a column from July of 2011, after requesting nominations from a wide variety of scientists, public officials, agency staff and others, a long list of nominees was reviewed and vetted by the National Academy of Sciences, and a committee of 13 individuals was selected. The list included scientists and engineers from universities and federal agencies across the country.

The committee met over a period of 18 months, reviewing all of the most recent data, reports and findings of researchers around the world. They invited other scientists to attend meetings and present their work and data, and then spent months digesting, discussing, debating before finally summarizing their evaluation in a report that was released in June of 2012 (http://www.nap.edu/catalog.php?record_id=13389). At 274 pages, the overall report is quite long, but the Summary is only eight pages and pretty straightforward and understandable. For those with an intense interest in the future rise in sea level along the west coast, or who happen to be an oceanfront home or business owner, it's worth a look.

Article 16

Sea-Level Rise
What Can We Expect?

Global sea level has been rising ever since the last Ice Age ended about 20,000 years ago and the planet began to thaw out. As the Earth and the ocean warmed, sea level rose for two main reasons: (1) ocean water expands as it warms; and (2) water from melting glaciers and ice sheets flowed into the ocean.

During the past 20,000 years of gradual warming, about 10,000,000 cubic miles of water has been added to the ocean from ice melt. That's a lot of water and it raised the level of the ocean about 350 feet around the world, gradually bringing the ocean right up to today's shoreline, and to the doorsteps of many homes. Humans had virtually no influence on climate or sea level for the first 19,900 years of this period, however. It was all natural.

Tide gages show that sea level rose globally about 7-8 inches over the 20th century, not a lot unless you happened to live within six inches of sea level in Bangladesh, on some very low-lying Pacific Islands, or perhaps below sea level in a place like The Netherlands, New Orleans or the Sacramento delta.

That's the past, however. California and most other coastal states and nations are concerned with what is likely to happen in the future. How high is sea level likely to rise and how fast? And that's what the National Research Council Committee evaluated for 18 months.

Satellite observations over the past 20 years tell us that the rate of global sea-level rise has nearly doubled over what it was during the previous century. In large part this is due to land ice melting at a more rapid rate, particularly in and around Antarctica and Greenland. There are still a lot of unknowns, however.

It has been said that prediction is really difficult, especially about the future. And sea-level rise predictions are no different.

There are several different ways that we can develop projections for future sea level: 1) extrapolate historic trends; 2) use climate models that incorporate a range of greenhouse gas emissions (carbon dioxide, nitrous oxide and methane); or 3) develop historical relationships between global temperature and past sea level, and then use estimates of future temperatures to predict future sea levels. The Committee used a combination of all of these approaches.

The West Coast Sea-Level Rise Committee concluded that based on all existing data, that we can expect global sea level to rise somewhere between 3 and 9 inches by 2030, compared to 2000 levels; rise 7-19 inches by 2050, and 20 to 55 inches by 2100. The farther we go out into the future, the greater the uncertainties become, simply because of the unknowns in things like global production of greenhouse gases. How much more coal, oil and gas will the U.S., China and India use in the decades ahead? No one knows the answer.

There is an analogy here with how much money you might need to save for retirement. If you are retiring in 10 or 15 years, you can probably make a reasonably good estimate. But if retirement is 25 or 35 years away (say by 2050), you have to deal with a whole lot more uncertainty: interest rates on your savings, the stock market and your investments, inflation rates, costs of living, on and on. The committee had similar challenges with the future of sea level.

While these global values listed above are useful to keep in mind, sea-level rise isn't the same everywhere, primarily because the land or coastline may be moving up or down. Large-scale tectonic forces, such as those that raised the flat terraces along the Santa Cruz coast, can drive vertical land motion. Because of the withdrawal of oil, gas or water from subsurface reservoirs or aquifers near the coast, or the compaction of sediments, some sections of coastline may actually be sinking. Huntington Beach is sinking and so are Venice and Mexico City.

So while the total volume of water in the ocean will determine the average global sea level, the rate of rise along any particular coastline is also affected by what that particular land area might be doing, whether rising, sinking or stable.

Lateral movement along the San Andreas Fault dominates the tectonics of the coastline from San Diego to Cape Mendocino in northern California. There is little vertical land motion so that projections for future sea level are similar to global values: 2-12 inches by 2030, 5-24 inches by 2050 and 17-66 inches by 2100.

Confidence in these projections is high for 2030, and perhaps 2050. But by 2100, due to the uncertainties mentioned previously and discussed in the report, we are only confident that the actual value will fall within the range given: 17 to 66 inches, or 1.5 to 5.5 feet. While that's a big range, even the midpoint of the range, 3.5 feet, spells trouble for many coastal areas. But the good news is that we have some time to adapt if we start right now. This could be called sober optimism.

PART VI

NATURAL DISASTERS

Article 1

Tsunamis - Should We Worry?

The word "tsunami" generates an emotional response, much like earthquake or shark. It's just one of those scary things that is beyond our control and that we never want to experience. The 230,000 people that died the day after Christmas in 2004 from the Indian Ocean tsunami was a cruel reminder of the forces that lurk offshore around the Pacific rim.

While most tsunamis result from subduction zone earthquakes in deep trenches, they can also occur where oceanic volcanoes erupt catastrophically or where large landslides run down steep underwater slopes. The biggest tsunamis that might ever occur come from asteroid impacts, but so far, no large ones have struck since humans have been on the scene. The 6-mile-wide asteroid that smashed into the Gulf of Mexico 65,000,000 years ago was traveling at 45,000 mph and created waves 100s of feet high that ran up a then inland sea as far as South Dakota. No worries though, asteroid impacts even one-hundredth that size hit Earth only once in about a million years on average.

With so many others things to worry about, do we also need to get stressed out about tsunamis here in Surf City? The short answer for me is that it's not up there on my list of top 10 concerns. Your odds of dying in a tsunami in Santa Cruz are far lower than virtually any other risk we all face daily: commuting over Highway 17, biking on Mission Street or riding a motorcycle on Highway 9.

On the other hand, tsunami hazards are something that city planners, public safety officials and structural engineers ought to be aware of and plan for. There have been seven tsunamis large enough to cause significant damage along the coast of California over the past 200 years. Over this time span, 16 lives have been lost; one of those was in Santa Cruz. On April 1, 1946, an older man was drowned while walking along Cowell Beach when water rose fifteen feet above normal quickly from a large earthquake in the Aleutian Trench off Alaska. Crescent City on the north coast was hit hard by a tsunami from the huge Good Friday Alaskan earthquake of 1964. Water levels rose 8 feet, and much of the low lying downtown area was inundated as waves washed 2000 feet inland, drowning 11 people and destroying 150 businesses. Water levels in the Santa Cruz harbor surged 11 feet, sinking a dredge and a 38-foot boat.

Following the 2004 Indian Ocean earthquake, concern was expressed about the risk from a tsunami generated in the "trench" in Monterey Bay. Fortunately the gash that cuts through the center of the bay is a submarine canyon, which is an undersea drainage system, and not a trench that generates large earthquakes. The San Gregorio-Hosgri Fault does slice through the ocean floor 10 miles offshore and it has the potential to generate magnitude 7 earthquakes. The ocean floor on either side of this fault however, is sliding north and south rather than up and down as needed to produce a tsunami. To the north lies the Cascadia Subduction Zone, however, perhaps one of the greatest tsunami threats to us here, but that story is yet to come.

Article 2

Subduction Zones, Great Earthquakes, and Tsunamis

On the evening of January 26, 1700, residents of several villages along the east coast of Japan recorded a series of tsunami waves washing up on their shoreline. Two hundred and sixty-five years later, as an oceanography graduate student, I made an interesting discovery that would take 20 more years to connect to the Japanese tsunami of 1700. I was studying sediment cores collected from an undersea canyon 150 miles off the Oregon Coast. This feature, Cascadia Channel, is much like Monterey Submarine Canyon and begins not far off the mouth of the Columbia River and extends for over a thousand miles across the deep-ocean floor.

What struck me as I opened up these cores of ancient sediment up to 30 feet long, was the very regular appearance of submarine mudflows deposits that had been left behind in this undersea canyon. Underwater mudflows, called turbidity currents, are very fluid, are driven by their greater density than seawater, and can flow hundreds of miles across the ocean floor. We have found similar deposits in Monterey Submarine Canyon and in other submarine canyons all over the world. These layers form when thick deposits of sand and mud brought to the ocean by rivers become unstable, begin to flow down slope, and are finally deposited many miles away in deep water.

By using carbon-14 to date the organic material contained in these individual 2 to 3 foot thick layers, as well as identifying volcanic glass in the sediments from the eruption of Mt. Mazama (now occupied by Crater Lake) about 7,700 years ago, I was able

Part VI: Natural Disasters 279

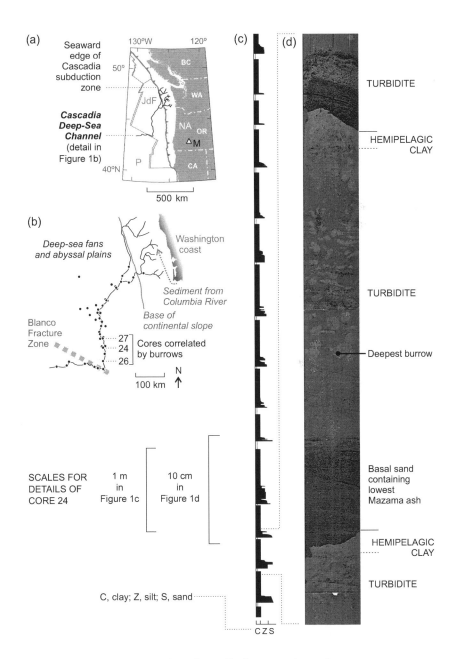

Turbidite from Cascadia deep sea channel.
(© 2011 Brian Atwater & Gary Griggs)

to determine that these massive mud flows happened about every 300-500 years. The question that plagued me at the time was what would initiate these huge underwater avalanches every several centuries? The two best ideas I could come up with at the time were either large storm waves or very large earthquakes. The year was 1966, however, and plate tectonics had not yet been born.

Within a few years, the global distribution of earthquakes, volcanoes, trenches and mountain ranges, as well as many other lines of evidence were recognized and assembled into a coherent picture of a mobile and fractured Earth. The Ring of Fire around the Pacific Basin was integrated into the concept of plate tectonics and we realized that the chains of volcanoes and trenches extending from New Zealand and Chile to the Aleutians were all part of a bigger picture of an Earth under stress. The deep-sea floor from Cape Mendocino to Vancouver Island was soon recognized as a subduction zone where an offshore oceanic plate slides beneath the coast of northern California, Oregon and Washington. This feature was given a name, the Cascadia Subduction Zone. We now know that the same type of plate motion that generated the 9.3 Indian Ocean earthquake in December 2006, as well as the March 2011 massive Japanese earthquake, and their associated tsunamis, also occurs along the Cascadia Subduction Zone. Twenty years after my initial discovery and the birth of plate tectonics, evidence for historic tsunamis and periodic great earthquakes began to be recognized in a number of estuaries and bays along the coast of Washington, Oregon and northern California. What does all this mean for Monterey Bay?

Article 3

Cascadia - A Sleeping Giant

Tsunami warning sign.
(© 2006 Gary Griggs)

Pounding El Niño storm waves during the winter of 1997-98 exposed an ancient drowned forest along the central Oregon shoreline. Over 200 tree stumps dating back to the time of Jesus became an instant tourist attraction. Through dendrochronology, or the study of tree rings, scientists were able to confirm that these spruce and cedar trees were submerged during a massive earthquake along the offshore Cascadia Subduction Zone about 300 years ago. Exposure to salt water killed but preserved the trees, and then beach sand and estuarine mud gradually buried them for over 300 years.

Geological observations combined with Carbon-14 dating and dendrochronology from sites along the coastlines of northern California, Oregon and Washington provide clear evidence for a great earthquake in 1700 that ruptured at least 600 miles of the Cascadia subduction zone producing a large tsunami. The sand deposited by this and other tsunamis in estuaries of the Pacific Northwest record seven big earthquakes in the past 2000 years, or one about every 333 years on average. It has now been over 300 years since the last great earthquake, so odds are gradually increasing for the occurrence of another very large event.

If the next big earthquake is anything like the last, we could expect 600 miles of rupture with up to 50 feet of slip. The coastline from Cape Mendocino to Vancouver Island would take the brunt of the tsunami, however, with wave heights of 20 feet or more expected. Waves would arrive within 30 minutes so warning time would be very short indeed. Most coastal communities are now aware of tsunami dangers and warning signs showing evacuation routes have been widely posted. For many low-lying communities, however, there quite frankly wouldn't be enough warning time to get very far from the shoreline. The coastline may likely be down dropped due to drag along the offshore fault, which would flood coastal development and inundate low-lying land.

Most tsunami energy from large subduction zone earthquakes moves perpendicular to the trend of coastline or the trench, or in this case, either directly on shore or out into the Pacific towards Japan. The shoreline of central and southern California would therefore be somewhat sheltered from large waves generated by a major Cascadia earthquake, which is nearly due north of us. While there is no historic record of what happened here during the huge earthquake of 1700, models or simulations of such an event indicate maximum wave heights of about 6 feet along the central coast.

Tsunamis generated by large earthquakes in the Aleutian trench in 1946 and 1964 raised water levels along the Santa Cruz coastline higher that those predicted from a large Cascadia event. Although we are about 1,500 miles farther away from the Aleutian trench, we are more directly in the path of Alaskan tsunamis.

On November 15, 2006 a tsunami from an 8.3 magnitude earthquake in the Kuril trench north of Japan raised water levels over 6 feet in the Crescent City harbor and left $700,000 in damage. The Santa Cruz harbor experienced three-foot surges of muddy water. A local tsunami generated by a large slump in the head of Monterey Submarine Canyon would provide the least warning time, ten minutes perhaps, and could produce significant run up. Evidence of large slumps exists in the canyon, but there is no record over at least the past 150 years of any tsunami generated just immediately offshore. So far, so good.

Boats in the Santa Cruz Harbor get tossed around in the tsunami from the March 2011 Japan earthquake.
(courtesy: Santa Cruz Port District)

Article 4

Tsunamis - To Worry or Not to Worry?

The earthquake and tsunami in Japan in March 2011, combined with a fascination many of have with giant waves, led me to pursue this topic a bit further. While the waves at Mavericks and Ghost Tree are wind driven, tsunamis are generated by large displacements of water, from huge ocean floor earthquakes, submarine volcanic eruptions, or massive landslides, either underwater or on land, that slide into the ocean, a bay, lake or reservoir. There are good historic examples of each of these scattered across the globe, but to date, the observable tsunamis we have experienced historically along the California coast have all been driven by distant earthquakes.

The world has experienced some devastating earthquakes over the past several years, with three of the six largest recorded over the last century all happening within the past five years: Sumatra 9.1 magnitude in 2006; Japan 9.0 in 2011; and Chile 8.8 in 2010. Rounding out the top six were: Chile 9.5 magnitude in 1960; Alaska 9.2 in 1964; and Kamchatka 9.0 in 1952. Interestingly, the world's greatest earthquakes occur at subduction zones or trenches, and most of these encircle the Pacific Ocean, so have the potential to generate tsunamis that would impact the coast of California.

The magnitude of an earthquake is directly related to the surface area of the plates or slabs that rupture. The larger the rupture area, the greater the amount of energy released. While the 6.9 magnitude Loma Prieta earthquake seemed large to all of us here in 1989 (and devastating locally), it was actually a moderate shock in the big picture of things seismic. The total area that ruptured in 1989 along the San Andreas Fault was only about 25 miles long.

In contrast, the great 9.1 Sumatra earthquake of December 26, 2004 took place where the Indo-Australian plate is descending beneath the Eurasian plate and ruptured the greatest fault length of any earthquake every recorded, spanning a distance of 900 miles! The energy released during an earthquake is proportional to the earthquake magnitude, which is based on a logarithmic scale. Each unit increase in magnitude corresponds to a release of approximately 30 times more energy.

The Sumatra earthquake and the massive 9.0 shock in Japan in 2011, therefore, released 30 times 30 or about 900 times more energy than Loma Prieta! Both the Sumatra and Japan earthquakes also displaced huge slabs of seafloor producing large pulses that propagated across the Indian and Pacific oceans, respectively. Waves along the coast of Indonesia in 2004 reached elevations 100 feet above sea level and led to the loss of perhaps 230,000 lives. Waves along the coast of Japan reached 128 feet in elevation, and while the number of casualties was considerably less, they still approached 18,000.

Five of the six largest earthquakes ever recorded on seismographs have occurred around the margins of the Pacific Basin, and these have been about as large as seismologists can imagine. We might conclude that we have already experienced the worst that can be expected in terms of tsunami impacts here in California, but there are some worse case conditions that might produce even greater impacts.

Over the past nearly 200 years or so of somewhat reliable news reporting in California there have been seven tsunamis that have been considered destructive. Over this period, 17 lives have been reported lost due to tsunamis, far less than the number of people killed by dog bites or bee stings. The earliest reported event was in 1812, when a tsunami caused by an earthquake in the Santa Barbara Channel generated large waves that washed ashore at Santa Barbara, and further west at Gaviota and Refugio canyons. Heights are not known for certain, but based on the scattered descriptions, the distances that the waves washed inland, and the damage left behind, they may have been 10 or 12 feet high.

A tsunami that appears to have been generated by an offshore earthquake hit the coast of Los Angeles, Orange and San Diego counties in August 1930. Maximum wave heights appear to have been ten feet or a little higher; many swimmers were rescued and one person is reported to have drowned along the Santa Monica shoreline.

Prior to 2011, the tsunamis that had been the most damaging to the California coast were both generated by large earthquakes in the Aleutian Trench off of Alaska, one on April Fool's Day in 1946 and the other on Good Friday in 1964. The 1946 tsunami had modest impacts from Noyo Harbor on the Mendocino Count coast in the north, where many boats broke from their moorings, to Santa Catalina Island in the south. Santa Cruz' only tsunami fatality, ever, took place from that tsunami, when an elderly man walking along the shoreline was drowned when the water level rose quickly to 10 feet above normal. In Half Moon Bay, which sits at a very low elevation, 14-foot waves reportedly washed a quarter mile inland, damaging houses, boats and docks.

Article 5

Lituya Bay
The Largest Wave Ever Witnessed

Lituya Bay after the tsunami of 1958.
(photo: D. J. Miller, US Geological Survey via Wikipedia Commons)

On the night of July 9, 1958, a 7.7 Magnitude earthquake along the Fairweather Fault in southeast Alaska shook loose about 40 million cubic yards of rock (4 million dump truck loads) high above the northeastern side of Lituya Bay. This huge mass of rock plunged from an elevation of about 3000 feet down into the bay. The impact of the rock and debris generated a local tsunami that washed 1720 feet up the ridge on the opposite side of the inlet.

For comparison, the Empire State Building with its antenna is 1470 feet high. The wave hit with such power that it swept completely over the spur of land on the opposite side of the inlet, removing all trees and vegetation from elevations as high as 1720 feet above sea level. Millions of trees were uprooted and swept away by the wave, the highest wave ever documented.

The wave then continued down the seven-mile length of Lituya Bay, ripping out or snapping off trees on either side of the bay at elevations up to 600 feet and then washed over a sand spit and into the Gulf of Alaska. The force of the wave stripped the soil off down to bedrock and snapped off large Spruce trees, some with trunks up to six feet in diameter.

There were three fishing boats anchored in Lituya Bay on the night the giant wave washed through. Orville Wagner and his wife Mickey were killed when their boat was sunk after being hit by the wave. Bill and Vivian Swanson, who were asleep in the Badger, and Howard Ulrich and his 7-year old son Junior, in the 38-ft Edrie, all survived. Both boats were anchored about a mile up Lituya Bay from the entrance.

Howard Ulrich reported hearing a deafening crash, resembling an explosion, at the head of the bay about 2.5 minutes after the earthquake was first felt. Based on Bill Swanson's description of the length of time it took the wave to reach his boat after overtopping Cenotaph Island in the middle of the bay, the wave may have been traveling up to 600 mph.

The violent motion of the waters from the earthquake awoke Howard Ulrich who watched the mountains shaking and clouds of dust coming from their peaks. After observing the chaos for about two minutes he noticed a gigantic wall of water coming down the inlet towards them, cutting a swath of trees along both shorelines. He estimated the wave as 50 to 75 feet high and very steep as it got closer. Finally realizing that he had to respond, he got a life jacket on his young son, started the engine but was unable to raise the anchor before the wave struck. He had steered the Edrie to face the wave directly and as she rose, the anchor chain snapped. The vessel with Howard and his son was carried toward and possibly over the south shore by the wave, and then toward the center of the bay by the backwash.

The water in the bay swashed back and forth for about 30 minutes and then became calm. After keeping the boat under control throughout this violent ordeal, Howard and Junior Ulrich powered out of Lituya Bay ay 11:00 pm that night.

The Swansons were also very fortunate. The Badger, still at anchor, was lifted up by the wave and carried completely over the sand spit at the entrance of the bay, stern first and riding the wave like a surfboard. Bill Swanson reported looking down at the top of the trees, estimated at about 80 feet tall, as they were carried over the spit. The wave broke and the boat hit the bottom and began taking on water. The Swansons abandoned their sinking boat, got into a small dingy and were fortunately rescued by another fishing boat two hours later.

This wasn't the first event to generate large waves in Lituya Bay, however. Ship logs of the famous French Explorer LaPerouse (who is credited with the discovery of the Bay in 1786) commented on the lack of trees and vegetation on the sides of the bay, "as though everything had been cut cleanly like with a razor blade". Other early explorers had also commented on successive lines of cut trees, indicative of other past large landslides and inundations.

Photographs of trim lines, where all of the trees were removed, taken from 1894 to 1929 show that at least one and possibly two waves occurred between 1854 and 1916. These trim lines were largely destroyed by a huge 1936 wave that uprooted and broke trees off as high as 500 feet around the bay. The 1958 wave, however, removed all of the previous evidence and was the largest wave yet. Because of the unique geologic and tectonic conditions of Lituya Bay, such giant waves will undoubtedly occur again in the future.

Article 6

Holding Back the Sea

King Canute reproving his courtiers and holding back the waves.
(painting by R. E. Pine, engraving by E. Holl, 1848,
via Wikipedia Commons)

A thousand years ago, when national boundaries were a bit different that today, there was a wise King who ruled both sides of the North Sea, including what today is occupied by Denmark, Norway, England, and Scotland. He was apparently quite well liked, had performed well in a number of wars, and also conducted himself gracefully in most kingly matters.

The great men and military officers who hung around King Canute were always praising him. "You are the greatest man that ever lived", one would say. Then another would exclaim, "O king! There can never be another man so mighty as you."

The king, however, was a wise man with much good sense, and he grew very tired of hearing such foolish words. One day he was hanging out at the beach, as he did from time to time, surrounded by his officers. They were praising him, as they were in the habit of doing, and he thought it might be a good opportunity to teach them a lesson. So he requested that they set his chair on the sand close to the edge of the water.

" Am I the greatest man in the world?" he asked. They all cried out, "There is no one as mighty as you". "Do all things obey me?" he asked. "Great Canute, there is nothing that dares to disobey you" they replied.

"Will the sea obey me?" he asked. By this time the tide was coming in and the waves were getting a bit closer. The foolish officers were afraid to say no and kept silent, all but one who exclaimed, "Command it, O king, and it will obey."

"Sea", Canute said with authority, "I command you to come no farther! Waves, do not dare to touch my feet!". But the tide came, just as it did every day. The water rose higher, coming up to the king's chair, and wet not only his feet, but also soaked his royal robe.

The officers were a little concerned by this and not sure what to expect next. With great drama, King Canute stood up, took off his crown and threw it on the sand declaring he would never wear it again. Hoping to teach all of those surrounding him a lesson, he reportedly said something like "All the inhabitants of the world should know that the power of kings is vain and trivial".

King Canute's story is often used to make the point of the futility of trying to hold back the sea. Well, the legislators in North Carolina apparently did not read the story. In October 2012 they approved a new law that just went into effect that attempts to regulate future sea-level rise.

The new law, signed by the governor, mandates that only the Division of Coastal Management will be allowed to put out an estimate of the rate of sea-level rise, and that these rates shall only be determined using historical data. Rates of sea-level rise may not include analysis of trends or scenarios of accelerated rates.

The bill was written in response to an estimate by the state's Coastal Resources Commission that sea level would rise by 39 inches by the end of the century, a significant increase over the past century. This prompted fears of more costly home insurance and accusations of anti-development alarmism among residents and developers in the state's Outer Banks region.

Tom Thompson, president of NC-20, a coastal development group and a key supporter of the law, said the science was flawed and that the Resources Commission failed to consider the economic consequences of the 39 inch rise: numerous new flood zones would have to be drawn, new waste treatment plants would have to be built, and roads would need to be elevated. This would cost the state hundreds of millions of dollars.

A geologist on the State Science Panel in North Carolina said every other state in the country is planning on three feet of sea level rise or more by the end of the century. Maine is preparing for a rise of up to two meters by 2100, Delaware is using 1.5 meters, Louisiana a meter, and California 1.4 meters.

One North Carolinian, writing in Scientific American, put it quite well; he stated that this legislation is tantamount to telling meteorologists "do not predict tomorrow's weather based on satellite images of a hurricane swirling offshore, moving towards us with 60 mph winds and ten inches of rain. Predict the weather based on the last two weeks of fair weather with gentle breezes towards the east. And don't use barometers and radar images. Use the Farmer's Almanac and what grandpa remembers."

Article 7

Superstorms and Storm Surges

In late October of 2102, a hurricane with the innocuous name of Sandy ravaged the Eastern seaboard of the U.S., battering the coastline, leaving cities flooded, and millions without power. In many areas, cell phone communication was down. Now that's a real disaster. The New Jersey coast was devastated and New York City was again at ground zero as one of the hardest hit regions.

The storm surge pushed water levels up to a record-breaking 13.88 feet above low tide level at Battery Park, on the southern tip of Manhattan. Water flooded subways, tunnels, streets and parking garages, and flowed into the construction site at Ground Zero like a flood. Joseph Lhota, the Chairman of the Metropolitan Transit Authority of New York City referred to it as the most devastating disaster the subway system has experienced in its 108-year history. 117 people died as a result of the superstorm, most from drowning.

We usually think of Atlantic City as a summer resort town, but it was transformed into a river on the evening of October 22, when high tide put 85 percent of the city underwater. Hundreds of feet of boardwalk washed away and many oceanfront homes were completely destroyed or seriously damaged; many were torn off their foundations and carried inland.

Over the past century, while the number of hurricane related deaths in the U.S. has decreased, the cost of damages has dramatically increased (Hurricanes Irene, 2011 >$10 billion; Wilma, 2005 >$23 billion; Ike 2008, >$30 billion; and Katrina, 2005, >$90 billion). Eight of the eleven most costly hurricanes in the U.S. occurred between 2004 and 2011. Damage from Sandy was over $68 billion.

The reduction in death toll in recent years is at least partly due to our improved ability to predict the landfall location, communicate to the public and evacuate at-risk populations. Increased property losses, however, reflect rapidly growing coastal populations, more construction in hazardous locations, and more expensive buildings. Atmospheric pressure, wind speed, height of storm surge, hurricane path, coastal topography and elevations, and intensity of development all affect the impact of any individual hurricane.

While the east coast was preparing for the arrival of what may have been the largest storm to ever hit the New York/New Jersey coastline, we were enjoying a warm sunny weekend. Their ocean backyard can be much more treacherous and unforgiving than ours along Monterey Bay.

One major difference between our two coasts is that we don't get many hurricanes; nor do we get the associated storm surges that typically wreak the most havoc, flooding subways, underpasses and neighborhoods and washing away homes, businesses, highways and beaches. Given a choice between several hurricanes (each accompanied by strong winds and storm surges) every year on the East Coast and the occasional earthquake and a big El Niño every 8 to 10 years here on the West Coast, I'll personally take my chances right here.

Storm surges are produced by the low atmospheric pressure and high winds that accompany hurricanes. Under normal conditions, the pressure of the atmosphere pushes down on the water surface, keeping the sea at a normal height.

But when a low-pressure disturbance moves over an area of ocean, the pressure on the sea surface is reduced and the height of the ocean rises. Coupled with 50 to 100 mile per hour winds, this results in seawater literally piling up against the shoreline, as it did on October 22, 2012, from the Carolinas to New England.

If the mound of seawater enters a bay, inlet, harbor or river channel, the water is funneled into a smaller area and can rise even higher. To make matters worse, Sandy arrived during a full moon, which produces the highest tides each month, elevating the sea surface even higher. In Manhattan, the peak of the storm surge coincided precisely with high tide. This was just plain bad luck.

The Gulf and Atlantic coasts have a long and deadly hurricane history, although there are some that stand out above others. The worst recorded natural disaster in U.S. history was a hurricane that made landfall with winds of 145 mph at Galveston, Texas on September 8, 1900. Galveston is built on a sandy barrier island with a maximum elevation of just over 8 feet. The storm surge reached a height of 15 feet, sweeping completely over the island, with most of the wooden homes and businesses simply floating away or crushed by waves. Between 6000 and 8000 people lost their lives.

Hurricane Camille barreled into the Mississippi and Louisiana coast in 1969, with a storm surge of over 24 feet, total losses of over $10 billion (2013 dollars) and 259 lives. Katrina in 2005 had a maximum storm surge of 25 feet at New Orleans and took over 1800 lives.

In 1983, during what was probably our most damaging California El Niño of the past century, sea levels were elevated above normal predicted tides along the central coast by about 12 to 24 inches for several months. Maximum storm surges along the California coast might be a foot or more, but compared to Atlantic or Gulf Coast hurricane surges of 10 to 25 feet, we live in paradise.

We cannot yet answer the specific question of whether climate change made Hurricane Sandy more likely to occur. What is already clear, however, is that climate change very likely made Sandy's impacts worse than they would otherwise have been.

Before power and phone lines had been restored, and the damage assessed, state officials were already requesting that the Corps of Engineers talk to them about the best way to rebuild the New Jersey shore. The federal government bailed out New York and New Jersey to the tune of $60 billion. Understandably there is a natural urge and compassionate desire to help out those who have lost homes and businesses. But within 14 months, two hurricanes hit the mid-Atlantic shoreline. Perhaps this is also is a good time to take a longer view and also think carefully about the future risks of rebuilding directly on the shoreline. While we can't stop these storms we can reduce their death and destruction by not rebuilding in low-lying areas that have been repeatedly damaged and eliminating federal incentives and subsidies to rebuild in the most hazardous areas.

Article 8

Weather on Steroids

Hurricane Sandy floods New Jersey taxi cab lot, October 31, 2012.
(© Dave S, licensed under CC BY-SA via flickr)

There was considerable speculation, writing and discussion immediately following Superstorm Sandy focused on how the impacts of this storm may have been amplified by climate change. Similar discussions and arguments were made about the 2012 summer drought that affected about two-thirds of the entire United States, as well as the large number of wild fires in the western states. And that drought has now continued through 2013 and 2014.

After our 100-degree weather in early October 2012, I wrote a column on the difference between weather and climate. Weather is short-term, day-to-day. Climate is long-term. But over time, there is no question that weather will be influenced by a changing climate. The weather extremes of all sorts we have been experiencing, whether highest average temperatures, driest summers, or smallest Arctic ice coverage, are all observations or trends we ought to be paying attention to.

The question about Superstorm Sandy and its relationship to global climate change echoes the speculation in the sports pages in 2012 about whether records like those of Lance Armstrong, Mark McGuire and Barry Bonds were aided by performance enhancing drugs. All three men were clearly exceptional athletes and may have had great careers anyway, but the steroids probably didn't hurt. We may now be experiencing weather on steroids.

There are several different ways climate change might have influenced the strength of Hurricane Sandy: through the effects of sea-level rise; through abnormally warm sea surface temperatures, which increases evaporation and can strengthen hurricanes; and possibly through an unusual weather pattern that some scientists believe bore the fingerprint of rapidly disappearing Arctic sea ice.

Two major factors affecting water levels are at work along the New York and New Jersey shorelines. First, rebounding of New England after the removal of the massive weight of Ice Age glaciers has caused the island of Manhattan itself and parts of New Jersey to slowly sink; think of a seesaw or teeter-totter with New England going up and Manhattan and New Jersey going down.

Second, at the same time, the oceans have been slowly rising. While the average global rise in sea level over the past century was 7-8 inches, the shoreline of Long Island, Manhattan and New Jersey experienced a rise of 10 to 16 inches. All this simply means that any storm surge will now reach further inland and reach higher elevations. While four to eight inches may not sound like a lot, where the land is very flat, this may allow seawater to flow for blocks. To the Metropolitan Transit Authority or Consolidated Edison, the main electric utility in Manhattan, each additional inch of sea level rise matters a great deal.

Climate change amps up other factors that contribute to hurricane strength as well. Higher ocean temperatures lead to increased evaporation and water vapor, which are both fuel for hurricanes. As the Earth's atmosphere has warmed, it retains more moisture, which is drawn into storms and is then dumped on us as rain.

Water temperatures off the East Coast were unusually warm during the summer of 2012 - September had the second highest global ocean and atmospheric temperatures on record - so much so that New England fisheries officials observed significant shifts northward in cold-water fish such as cod. Sea surface temperatures off the Carolinas and Mid-Atlantic remained warm into the fall, offering a perfect energy source for Hurricane Sandy as it moved northward from the Caribbean.

Some scientists, and I realize that there are a lot of scientists out there, believe that about 1°F out of the 5°F East Coast water temperature anomaly may have been due to man-made global warming. Warmer seas provide more water vapor for storms to tap into; this water vapor can later be wrung out as heavy rainfall, resulting in enhanced flooding.

The most damaging aspect of the storm was the massive surge that struck the coastline from Maryland to Massachusetts. The added 10 to 16 inches to the sea level rise of the last 100 years and gave the surge a higher launching pad than it would have had a century ago, making it more damaging than it otherwise would have been. The impacts of storm surges are only going to get worse as sea level continues to rise as a result of warming ocean waters and melting polar ice sheets and glaciers.

The storm surge at The Battery in Lower Manhattan was the highest ever recorded at that location. It surpassed even the most pessimistic forecasts, with the maximum water level reaching 13.88 feet above the average of the daily lowest low tide of the month, known as Mean Lower Low Water. The result was a tide that took water from New York harbor, the Hudson River, and the East river and put it right into downtown New York City.

Sandy was also somewhat unprecedented in its pathway. Hurricanes coming up the Atlantic seaboard almost always drift to the northeast and dissipate their energy at sea. Sandy made nearly a 90-degree turn towards the coastline. Why?

Hurricane Sandy crashed into cold air moving south from Canada. This collision increased the storm's energy level and extended its geographical reach. An atmospheric pattern above the Arctic Ocean, known as a blocking high, pushed that cold air into the path of the hurricane. Two climate scientists from Cornell University provided evidence earlier this year that Arctic ice melt linked to global climate change was contributing to the very atmospheric pattern that sent the frigid air down across Canada and into the eastern U.S. All of this put the National Weather Service forecasters in the odd position of having to issue snow advisories for a tropical-hurricane system.

Hurricane Sandy damage on the Jersey Shore, Nov. 2, 2012.
(photo: Jim Greenhill, National Guard, licensed under CC BY-SA via flickr)

Article 9

Perspectives on Disaster

Downtown Sacramento during the Great Flood, 1862
(courtesy: California State Library)

While the healthiest among us may live to be 100 or a few years more, in the large geological scheme of things, this is really just a blink of an eye. The Earth is about 4.5 billion years old, give or take a few million years. Depending upon your definition of "human", we have only been around for about 200,000 years.

To put our human existence into some perspective, if the entire 4.5 billion years of Earth history was compressed into a single year, human beings would have appeared at 11:36 pm on December 31st. The entire period of human occupancy of the Earth would have only covered the last 24 minutes of its one-year long history.

To reduce our geological significance a bit more - for the sake of argument - let's say that earliest human civilization began about

8,000 years ago. In our 365-day history of the Earth, this corresponds to 2 minutes before midnight on the last day of the year. I don't think anyone would disagree that we have had a disproportionate impact on the face of the Earth relative to our final two-minute presence.

Civilization's earliest written records first appeared about 3500 BCE in the Middle East. China began recording earthquakes about 3000 years ago, simply because they suffered repeatedly from some of the world's most damaging shocks. They also have a record of great floods that extends back about 2,000 years.

The longer our observations or records extend back in time, the more complete and accurate a picture we have of what we might reasonably expect to experience in the future. With natural disasters, the old proverb - what you don't know won't hurt you - really doesn't hold true. Whether we are building a power plant, a shopping center, a school or our dream home, the more we know about past events in the neighborhood, the better.

In California we are hindered somewhat by our short history. Our written disaster accounts only extend back about 200 years to the time when Europeans first began to occupy Alta California. But practically speaking, until about 1850 or so, records in the Monterey Bay area were pretty sketchy.

A summary of the earliest flooding accounts in Santa Cruz, for example, from the old Pacific Sentinel newspaper of 1871 states: "A flood occurred in the year 1822, when the water covered all the lowlands and rose to a greater height than ever before. The next memorable flood occurred in 1832, the water not reaching the extreme heights of ten years before, but still covering the lowlands to a great extent. The pioneer settlers of California remember distinctly the flood of 1852, and it is not necessary to dwell upon the particulars."

In 1889, J.M. Guinn wrote "History of California Floods and Drought", which provides an interesting historical perspective: "If there is one characteristic of his State, of which the true Californian is prouder than another, it is its climate. With his tables of mean temperature and records of cloudless days and gentle sunshine, he is prepared to prove that California has the most glorious climate in the world."

"For the first fifty years after the settlement of California the weather reports are very meager....although the state of the weather was undoubtedly a topic of deep interest to the pastoral people of California...with their cattle... and their flocks and herds spread over the plains, an abundant rainfall meant prosperity; a dry season, death to their flocks and consequent poverty....A flood might be a temporary evil, but like the overflow of the Nile, a year of plenty always followed."

Beginning on Christmas Eve, 1861, and continuing into early the next year, a series of storms lasting 45 days struck California, Oregon, and Nevada and produced the largest floods in recorded history for much of the state. The severe flooding turned the Sacramento Valley into an inland sea, forcing the State Capital to be moved from Sacramento to San Francisco for a time, and requiring Governor elect Leland Stanford to take a rowboat to his inauguration. This is in striking contrast to the drought years of 2012, 2013 and 2014.

William Brewer, author of "Up and Down California," wrote on January 19, 1862, "The great central valley of the state is under water - the Sacramento and San Joaquin valleys - a region 250 to 300 miles long and an average of at least twenty miles wide, or probably three to three and a half millions of acres!" The storms reportedly wiped out nearly a third of the taxable land in California, leaving the state bankrupt.

The 1861-62 series of storms were probably the largest and longest California storms on record, and the flooding of the Central Valley the most widespread in the state's short history. However, geological evidence suggests that earlier, prehistoric floods were likely even larger. There is no reason to believe that such extreme storms could not happen again, and predictions with a warming ocean are for more evaporation, and subsequently, more winter rainfall and more flooding.

So relative to China, Japan or parts of the Middle East, our record of past natural disasters, whether floods, earthquakes, droughts, tsunamis or other calamities, is really quite short. Nonetheless, we ought to pay attention to the history we have recorded, and perhaps a lot more attention now that our climate is changing and the prognosis is not particularly encouraging for California in general, or the Monterey Bay region in particular.

PART VII

MARINE LIFE

Article 1

Algal Blooms - Good and Bad

Algal blooms can be beneficial but also have a darker side. The pastures of the sea are the microscopic floating plants called phytoplankton. They serve an essential role as the base of the ocean food chain, but they also have a darker side that has come to light in recent years. While most phytoplankton are benign, some produce toxins that can have devastating effects on humans and wildlife.

The Native Americans who originally lived along our shoreline took advantage of the abundant fish and shellfish in the coastal waters, but learned that at certain times of the year the mussels were dangerous to eat. Legend has it that these early inhabitants posted sentries along the shoreline to warn against harvesting contaminated shellfish at dangerous times. Today, health officials closely monitor commercial shellfish harvests and post signs to warn those who might be looking for dinner on the rocks.

Massive blooms of phytoplankton often occur in Monterey Bay in late spring and summer, sometimes turning the water a reddish color when certain kinds of algae are present. Most of these algal blooms are harmless. Sometimes, however, certain species that produce potent toxins proliferate for reasons we don't completely understand. The result is a "harmful algal bloom," a better term than the misleading "red tide," since many harmful blooms aren't red, and many reddish blooms aren't harmful.

The toxic algae may be filtered out of the water during feeding by shellfish or consumed by tiny krill or small fish such as anchovies and sardines. Most fish and shellfish don't seem to suffer any obvious effects from eating the toxic algae, however. Unfortunately, the toxins get passed up the food chain. The real trouble starts when marine mammals, birds or people eat the contaminated fish or shellfish. Paralytic shellfish poisoning, amnesic shellfish poisoning and diarrhetic shellfish poisoning are serious diseases caused by different algal toxins. The toxins can affect the nervous system, causing partial paralysis, loss of balance and even death. On the Central Coast, these effects have been seen in seals, sea lions, pelicans and other seabirds.

A famous incident of seabird poisoning occurred in the summer of 1961, when a large flock of sooty shearwaters, fresh from a feast of anchovies, collided with coastal structures, windows, and cars from Pleasure Point to Rio del Mar. Residents were awakened in the middle of the night by birds slamming against their homes, and in the morning their yards and streets were littered with dying and confused birds and the smell of dead fish.

Alfred Hitchcock, who had a home in Scotts Valley at that time, apparently read about this event in the Sentinel and used it as the inspiration for his film of avian malice, "The Birds." The reason for this event remained unknown for more than 25 years until it was discovered by marine scientists at UC Santa Cruz who analyzing preserved samples from that event that the birds had been affected by domoic acid, a toxin produced by a particular type of phytoplankton.

Harmful algal blooms are naturally occurring events, but they do appear to be increasing in intensity and frequency. This may be due to natural cycles, but there is growing evidence that human activity may also be causing blooms to occur more frequently, be larger and last longer. One factor may be increased nutrient availability from sewage discharge or fertilizer runoff from agriculture. Pinto Lake in the south county has been recognized in recent years as a hot spot for dense concentrations of microcystin, a cyanobacteria, also known as a blue-green algae. Runoff from the lake finds its way into the Pajaro River and ultimately to the shoreline. As of 2010 the deaths of 21 California sea otters had been linked to this toxin, many of these near river mouths or harbors, and 17 of these from Monterey Bay. Necropsies of these otters have revealed acute liver failure or damage to other tissues, and also tested positive for microcystin. The otter diets include mussels, clams, crabs, and other marine invertebrates, which all can concentrate toxins from this blue-green algae to harmful levels.

ARTICLE 2

Global Migrations - Sooty Shearwaters

In Alfred Hictchcock's film "The Birds", the sooty shearwaters that were poisoned by the toxins from the anchovies they had gorged on, appear in the summer months along our shoreline in very large numbers. Many locals have probably noticed the huge black clouds of these birds out over the bay, which can number in the tens of thousands, flying just above the ocean surface and diving for small fish. Because many marine birds and mammals, whales and elephant seals, for example, are only temporary seasonal residents along our coast, biologists who study these animals have long wondered where do they go when they aren't here? The larger animals, like the grey whales, are pretty hard to miss as they swim just offshore. Their seasonal migration between summer feeding grounds in the Bering Sea and their winter calving area in the lagoons of Baja California have been known for decades.

The small sooty shearwaters, which are about the size of a large pigeon, posed greater tracking problems. Researchers at the University of California Santa Cruz Long Marine Laboratory, partnering with colleagues in New Zealand, France and Hawaii, figured out a way to track the bird's migrations and personal lives. Working with engineers, these scientists have developed increasingly smaller and more advanced electronic tags for all sorts of marine animals. The tags developed for the shearwaters are attached as leg bands in New Zealand where the birds nest in burrows. These tiny instruments are able to keep track of light levels and temperature over time, which provide a record of the latitude where the birds are on any particular day. These small instruments also record pressure, which indicates diving depths as the birds enter the water to feed.

All of this flight information is recorded by the tags and can be recovered when these migrating birds return months later to their nesting areas in the southern hemisphere.

What came as a big surprise several years ago when the data were first downloaded from these miniature recorders, was that these birds underwent the longest recorded migrations of any animal ever tracked! From their New Zealand nesting grounds, these birds leave in April and follow global wind patterns, heading first eastward, and then following the trade winds to the north Pacific where they feed throughout the summer months. The sooty shearwaters head to one of three different but highly productive feeding grounds, the western Pacific off Japan, the Gulf of Alaska, and the waters of the California Current. They dine on fish, squid and krill for several months and then turn around and complete a figure 8 path, flying all the way back to the southern hemisphere in a southwesterly direction, continuing their pursuit of an "endless summer" of seafood.

The most amazing part of this story is that these birds, weighing less than two pounds, complete a total migration of 40,000 miles over a period of about six months, equivalent to flying completely around the world at the equator one and a half times! On their best days they fly over 600 miles, taking advantage of the trade winds on both their northerly and their southerly migrations. Equally impressive are the dive records for the shearwaters. Not only do they fly 40,000 miles to feed, but the pressure recorders in their electronic tags indicate that they can dive over 200 feet below the surface searching for food. Sooty shearwater populations today are declining, with researchers concluding that because they operate on a truly global scale, that they may be serving as an important indicator of climate change and ocean health.

Part VII: Marine Life 311

Article 3

Tracking Marine Mammals

Satellite and data tag on female elephant seal.
(© 2012 Dan Costa)

Scientists opened the door to a marine world we had never seen before when they began to place electronic monitoring tags on marine mammals, large fish, sharks, turtles and seabirds. For the first time, they could follow the migration routes of these animals and track their movements above and below the sea surface.

Sophisticated tagging technology is central to a project called Tagging of Pacific Predators (TOPP), which began about 10 years ago and expanded to involve dozens of scientists around the world, including groups at Long Marine Laboratory, Hopkins Marine Station, the Monterey Bay Aquarium and NOAA's Pacific Fisheries Environmental Lab. Although after 10 years, TOPP evolved into the Global Tagging of Pacific Predators, the program successfully deployed over 4,300 electronic tags on 23 different species of marine animals throughout the Pacific Ocean.

Their discoveries include the record-breaking 40,000-mile annual migrations of sooty shearwaters, a Pacific Ocean endless summer flight spanning northern and southern hemispheres. By tagging 179 different great white sharks off the central California coast, researchers have learned that a small group heads toward the Hawaiian Islands and another gathers in an area 1,300 miles offshore, roughly midway between California and Hawaii, that has been dubbed the "white shark café". Exactly what they do in the "café" remains unknown, although feeding and mating seem high on the list of possible activities.

Only recently have electronic tags become small and light enough to be worn by a small bird like the sooty shearwater. The earliest electronic tags were so big they could only be used on large animals such as seals, sharks and whales.

UCSC biologist and TOPP investigator Daniel Costa began using electronic data recorders to study the diving behavior of elephant seals in 1983. Those studies yielded astonishing results--elephant seals routinely dive to depths of 2,000 feet and can go as deep as one mile beneath the surface on dives that can last an hour or longer. If you don't think this is an amazing feat for an air-breathing mammal, try sticking your head under water and holding your breath for two minutes.

More recent studies show that elephant seals from Año Nuevo Island swim thousands of miles offshore into the Gulf of Alaska each year, in some cases as far as the Aleutian Islands, to feed in ocean "hot spots" where productivity is high and food is abundant. Repeat tagging of the same animals shows that they follow nearly identical paths year after year, across thousands of miles of ocean with no signs or maps to guide them.

The miniaturization of electronic components and sensors, plus advances in satellite technology, has led to dramatic improvements in the tags. Data loggers work like miniature flight recorders attached to an animal to record location, time, diving depths, heart rates and, more recently, water temperature and salinity. The information collected can be recorded and stored in the tag for later recovery, or it can be transmitted through satellite links to scientists in labs around the world.

This technology is a far cry from the first simple tags developed a century ago to track fish migrations. Those worked much like a note in a bottle - scientists knew where and when they tagged a fish and where and when it was caught, but nothing in between. The GTOPP website (www.gtopp.org) allows you to sit in the comfort of your easy chair and follow these scientists and the migrations of the animals they have tagged all over the Pacific Ocean.

Each new tag deployed on a marine animal provides a glimpse into its life at sea. The challenge for scientists is to figure out why these animals follow the routes they do, how they know where they are going, and what they do when they get there.

Article 4

High-Tech Ocean Observations

The small electronic tags we have now developed and placed on migrating birds and the larger data recorders we have used for elephant seals, sharks and whales, are just two ways in which the high-tech world has invaded the ocean, allowing ocean scientists to explore, observe and monitor the oceans in ways that weren't even imagined a decade ago. Any of you sitting in front of your home computer can go on-line, 24 hours a day, and see what the ocean temperature is off Florida, what surf conditions are in Mexico, or which way the surface currents are flowing off of San Francisco.

While we have been making these types of observations for many years, the time and effort involved in the past was enormous. Until the present generation of satellites began orbiting the Earth, we needed to send ships out to collect most of the information we desired on ocean conditions; but much of this has now changed. Ships are still necessary for sampling the water beneath the surface and for collecting marine life, or sediments from the ocean floor. But you no longer have to go 1000 miles out to sea and suffer days of seasickness to study the ocean. Vast amounts of information on the oceans are now collected remotely in real-time on the ocean surface and also on the sea floor, not only from satellites, but also from shore-based observation systems, from remotely operated vehicles, anchored moorings, and also from free floating instruments..

The Monterey Bay Aquarium Research Institute (MBARI) several years ago placed what might be one of the world's longest extension cords across the seafloor, linking the deep ocean floor beyond Monterey Bay with the Institute 23 miles away in Moss Landing. This cable provides power and also data transmission for instruments that can be placed on the seafloor many miles offshore. Scientists from any institution can now use this power supply and data link to support instruments that can monitor earthquakes, measure deep ocean water temperature and chemistry, as well as transmit high definition video over extended periods of time in locations where we were formerly limited by the life of a battery pack.

Measuring ocean currents used to require either going out to sea and putting drifting instruments in the water or setting up recording current meters in locations of particular interest. But these observations were very local, and only provided data for the area in the immediate vicinity of the instrument or for the period of time that the ship was able to track drifters and record their positions. Not long ago, someone got the brilliant idea that there may be a better way to do this, as is often the case with many things we have done the same way for years. A shore-based system for observing ocean currents using radar on a continual basis was developed. A set of about 40 coastal stations was installed several years ago that spanned the entire coast of California and that uses high frequency radar to directly measure the direction and speed of coastal currents (www.cencoos.org).

This is all part of a regional California ocean observing system (CeNCOOS-the Central and Northern California Ocean Observing System) that is focused on collecting information about the coastal ocean and making it available to the public, user groups and government agencies for practical application. The travel paths of oil or sewage spills and probable locations of missing vessels, or the occurrence and extent of harmful algal blooms are just several examples of how new technology is helping us observe the ocean nearly continuously, and these observations can provide us with the information that allows us to make informed decisions on how to best respond to the particular incident or problem we may be facing.

ARTICLE 5

Reading a Salmon's Ear Bone

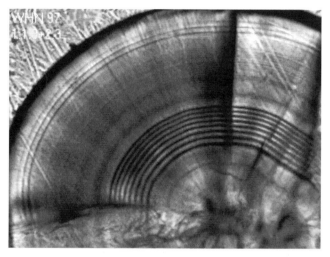

Central Valley Chinook otolith.
(courtesy: US Fish & Wildlife Service)

Fish, like humans, have ear bones, which are called otoliths. They're used by fish as sensory organs, playing a role in hearing and also balance. These small structures are made up of concentric layers of calcium carbonate and a protein material and grow throughout a fish's life. The accretion rate varies with growth of the fish - often slower in winter and faster in summer - resulting in the appearance of rings that resemble tree rings. By counting the rings, it is possible to determine the age of the fish in years, exactly like we would do with the rings in a log or tree stump. In most fish species, the accretion of calcium carbonate and the protein also alternates on a daily cycle, which makes it possible to determine fish age in days. Knowing the age of a fish and how fast it grows are important for understanding questions like how long particular fish species spend as juveniles, and what the population structure is. All of this information is in turn important for designing appropriate fisheries management policies.

The composition or chemistry of the individual layers of a fish otolith is also now proving extremely useful in providing us with information about the lives of fish such as salmon - questions like where they are born, how long they live in fresh water before they migrate out to sea, and how fast they grow. Much like the rings of a tree, otolith layers are deposited as a fish grows, with each layer recording the components of the chemical environment in which the fish was swimming on that particular day, week or month. Analyzing the composition or the trace elements within an ear bone allows us to figure out where a fish spent its life and how long it lived in each neighborhood.

What we need is a tool that can precisely analyze the composition of these very, very thin daily layers in a fish's ear bone. The instrument has now been developed to do just that, and has a name you will probably not soon forget, a laser ablation inductively coupled plasma mass spectrometer.

One of the most useful trace elements being analyzed is strontium, simply because it is chemically similar to calcium, so is easily taken up in the calcium carbonate of individual layers. Strontium comes in two different isotopes or forms, strontium-87 and strontium-86, and the ratio of these two isotopes in the waters of a stream is related to the rocks the streams flow through. Strontium ratios, therefore, are unique to individual watersheds, and their signature in a fish ear bone acts as a natural population marker, sort of a geographic birth certificate for a fish.

The Central Valley Chinook salmon formerly made up 90% - or about $60 million worth - of California's ocean harvested salmon, but the populations went through a disastrous decline for several years, which forced the closure of the commercial salmon season. Knowing something about the life histories of these fish and where they come from may help us resolve some of the problems with their declining populations.

Due in large part to more favorable oceanic conditions, the Chinook salmon rebounded in 2013 with a catch value of $22.8 million, number 3 in California's fishery.

By measuring the strontium-87 and strontium-86 ratios in the ear bones from salmon in the rivers of the Central Valley, and also from fish caught offshore in the adjacent ocean, scientists can tell with 95% accuracy not only where the fish was hatched, but also how long it lived in fresh water before it reached San Francisco Bay, and then when it entered the ocean. With this information we can determine how many of the ocean salmon had their origins in specific rivers (or hatcheries) in the Central Valley.

What came as a surprise when the data from the fish ear bones were first evaluated was that only 10% of the fall-run Chinook salmon were wild fish and the other 90% were hatchery fish. The role of hatcheries in the management of salmon populations has been a controversial issue and this high tech otolith research indicates that the decline in the natural salmon population has been masked by the larger numbers of artificially propagated or hatchery fish. Salmon have been a major fishery for California for many decades but for a complex combination of reasons, some natural and many human, the populations are in very serious decline and only occupy a small portion of their original range. Deciphering the tape recorders of their lives embedded in their ear bones may help us bring these fish back from the edge.

Article 6

Calamari-Small, Large, and Extra Large

Humboldt Squid.
(© 2013 Fish Guy, licensed under CC BY-SA 3.0
via Wikipedia Commons)

Squid come in three different sizes, and until fairly recently, all we saw were the smallest ones on our dinner plates. The small market squid has been near the top of the charts in recent years, being the largest California fishery by weight, 115,000 tons in

2013, 63.4% of the entire commercial fish catch, although most of this gets exported. Market squid were number 2 in value at $73.7 million, 28.7% of the entire catch, with Dungeness crab being the most valuable at $87 million. These little calamari, up to a foot long, have about a 12-14 month life cycle such that the entire stock turns over every year.

They are terminal spawners, meaning they spawn at the end of their yearly life cycle. Females generally produce 20 to 30 egg capsules with each capsule containing 200 to 300 eggs. So a successful female can produce 4000 to 9000 individual eggs, and very few animals eat the eggs. The market squid migrate in enormous schools throughout the eastern Pacific from the waters of southeast Alaska to Mexico. Most of the squid are caught off California, however, using very bright lights and nets at night. Squid are apparently like moths.

At the other end of the size spectrum is the Giant Squid, the largest invertebrate on Earth. The biggest ever found was about 60 feet long, and weighed nearly a ton. These big guys are very rarely observed, however, because they normally live in deep water. Like their small relatives, they have eight arms and two longer feeding tentacles that help them grab food. Curiously, their eyes are the largest of any animal around, up to 10 inches across. In 2004, researchers in Japan got lucky and took the first underwater images of a giant squid. Two years later, other Japanese scientists caught a 24-foot long female and brought her to the surface.

In June 2008, a research vessel on a shark tagging expedition 20 miles off Santa Cruz found the partial remains of a giant squid floating at the ocean surface. The tentacles were as thick as a human leg although much of the body was missing. The squid was taken back to Long Marine Lab where interested scientists performed a necropsy on the remaining body parts.

And now the somewhat smaller Humboldt squid has entered the bay and taken up residence. Still, at five to six feet long and 100 pounds they are known in Mexico as Diablo Rojo, or red devil, for their color and aggressive behavior. This carnivorous calamari has been reported in large numbers in recent years off San Diego and the Orange County coast to the north. While normally living at depths of 600 to 2000 feet, research indicates they appear to have established a year-round population off California, often at depths of 300 to 600 feet. There now has been frequent diver contact as shallow as 60 to 80 feet. Its not clear how many squid are offshore but based on their school size elsewhere, they could number in the hundreds or thousands. Their razor sharp beaks and long barbed tentacles have brought terror to scuba divers and close encounters with these devils appear to be increasing. Their growing numbers migrating northwards may be due to warming waters, or possibly a shortage of food in their normal territory to the south, or a decline in natural predators.

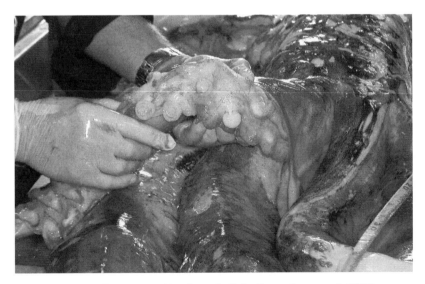

Remains of a giant squid collected off the Santa Cruz coast, 2008.
(courtesy National Marine Fisheries Service)

ARTICLE 7

Ocean Economics

While its difficult to put a dollar value on many things, economists are pretty adept and creative at figuring out ways to measure things in economic terms. If California were a country, our overall economy (the total value of goods and services produced by the state) in 2013 would rank us 8th in the world, after the United States, China, Japan, Germany, France, the United Kingdom and Brazil; going further down the list of nations of the world we would be ahead of Russia, Italy and India. Let's face it, despite all of our problems, all the things we complain and whine about, we are an amazingly creative and productive state.

A relatively recent effort, the National Ocean Economic Project, determined that California has the nation's largest ocean economy, which totaled about $46 billion in 2005. For comparison, agriculture produces about a $37 billion annually and it generates an additional $100 billion in related economic activity.

In order to make sense of something as complex and large as ocean business in some understandable way, the 2005 study divided the state's ocean economy into six major components:
- Construction - all marine related construction
- Living resources - commercial and recreational fishing, seafood markets and processing, and aquaculture or fish farming
- Minerals and energy - oil and gas exploration and production; sand and gravel extraction;
- Ship and boat building and repair
- Tourism and recreation - amusement and recreation; eating, drinking and lodging; marinas, RV parks and camping;

surfing, sailing, and other sporting activities; zoos and aquaria
- Transportation - Cargo shipping, marine transportation and services, search and navigation equipment, and warehousing.

Tourism/recreation and transportation made up about 93% of California's total ocean economy in 2005, the last year the data were collected. Living marine resources, all of those fishing related jobs and income, constitute only about 2.5 percent of the total. Even more surprising, the mineral and energy sectors, mostly southern California offshore oil and gas, made up less than 1% of our total ocean economy! This would be a totally different percentage for the Gulf Coast states with over 4000 oil drilling and production platforms operating in the Gulf of Mexico. A major difference is that over the long haul, oil and gas are nonrenewable and some day will be gone. Tourism and recreation, key elements of California's coastal ocean economy, can be sustained forever if we maintain a healthy coastal ocean.

An increasingly important component of our Monterey Bay region is marine research and education. There are presently about 25 individual ocean research and/or educational institutions, government agencies or programs located between Santa Cruz and Pacific Grove, with three hubs or centers: Santa Cruz, Moss Landing, and the Monterey Peninsula. These range from 1] research institutions like the Institute of Marine Sciences and Long Marine Laboratory at UCSC, Moss Landing Marine Laboratories, the Monterey Bay Aquarium Research Institute, the Naval Postgraduate School, Hopkins Marine Station, and CSU Monterey Bay; 2] state and federal ocean agency laboratories and programs such as the United States Geological Survey, the National Marine Fisheries Service, the Monterey Bay National Marine Sanctuary, Fleet Numerical Meteorology and Oceanography Center and the Naval Research Laboratory, to name a few. We also have 3] public education facilities like the Monterey Bay Aquarium, the Seymour Marine Discovery Center, and the Sanctuary's Exploration Center.

These institutions employ over 2300 scientists and support staff and have a total annual budget of about $314 million. Funding comes from a combination of federal agencies, state agencies, private foundations, industry and the public. Over the past two decades the Monterey Bay region has become recognized as a national and international center for ocean research and education, which continues to attract additional programs and scientists to the area. We have an exceptional coastal ocean environment for research and education, from Monterey Submarine Canyon to a biologically rich and productive coastal zone, which have attracted one of the largest and most diverse groups of marine scientists in the world.

Article 8

Sharks - Who's Attacking Whom?

On a beautiful warm fall afternoon during my freshman year at UC Santa Barbara, I returned to our dorm room to find my roommate huddled under a blanket, visibly shaking. I asked him what was going on?

I liked Clark and was concerned about what might have happened to him. He and I had become good friends and were the only guys in our small, former Marine Corps barracks dormitory that surfed. Somehow, the questionnaire they had given to freshman that they used to try and match up roommates had worked out in our case.

The year was 1961, and foam surfboards were still pretty new. Clark had an old, balsa wood board; one that had seen the ravages of time. When he was out on the board, it was about half submerged from getting waterlogged over the years. And that was part of the explanation of why he was in bed shaking on a warm September afternoon.

I slowly coaxed the story out of him. He had been out paddling in the kelp beds off Isla Vista, just west of the campus. Clark was also a scuba diver and had an interest in anything that moved in the ocean. He was a few hundred yards offshore that afternoon when he saw a dark shape that he thought was a seal or sea lion. He eagerly paddled over to investigate. When he looked down, what he saw almost sent him into shock. Instead of a seal, he saw a large shark directly beneath him. And his board, partly submerged, was just grazing the top of the shark's body.

Some of the advice they used to provide for downed pilots who find themselves in shark-infested waters is to do a relaxed breaststroke towards the beach, so as not to agitate the water and attract the attention of the shark. My roommate made it to the beach, but from his condition when I saw him, I don't think he was calm or relaxed as he paddled back with his old, partially submerged, balsa board.

In all of my years in the ocean, I never had any experience remotely close to what my roommate had that afternoon. But like 99% of us, encountering a shark in the water is one of those things I would just as soon avoid. A shark attack to a surfer, paddler, kayaker or swimmer, is something that we just instinctively fear. Horror fiction and films such as Jaws have tapped into this fear while undoubtedly feeding it further.

While all of the roughly 375 shark species worldwide are carnivores, none has any personal agenda against people, although occasionally sharks will mistake a person for prey. Contrary to popular belief, only a few types of sharks are dangerous to humans. Of all of the species of sharks, only four have been involved in a significant number of fatal unprovoked attacks on humans: the great white, tiger, bull and oceanic white tip.

So what are the odds? How often to people get attacked by sharks? There are lots of interesting statistics out there to provide some perspective, and hopefully some peace of mind. Over the past 85 years of so of reasonably accurate records, there have been about 110 shark attacks in California history resulting in 10 fatalities. That's about one fatal attack every 10 years for the literally millions of people who use California's beaches, and only one of those has been in Monterey Bay. One in over a century. You're more than 300 times more likely to be killed by dog bite than a shark, and over 500 times more likely to be killed by a wasp, hornet or bee sting. I hope that makes everyone feel better.

If you really want to get worried about your daily activities, just consider this: in contrast to the occasional dog bite or bee sting, motor vehicle related deaths - whether you are in the car, on a motorcycle, bicycle or a pedestrian - account for nearly 45,000 deaths across the country each year. That's five people every hour, 24 hours a day, 365 days a year!

How about cell phones? This is something we have some control over. Texting while driving makes a driver 23 times more likely to crash, and talking on a cell phone, 4 times as likely to have an accident. 46% of drivers under 18 admitted to texting while driving and 21% of distracted teen drivers involved in fatal accidents were reported to be distracted. In 2012, 3,328 people were killed in distraction-related accidents, and 421,000 were injured.

But, I digress. Let's get back into our ocean backyard. Although I may have convinced you that just about anything you do every day is far more hazardous than being attacked by a shark, it does happen from time to time, but very rarely.

Not all of the numbers reported agree, but to provide you with some picture, the United States has had more reported shark attacks than any other country, on average about 16 attacks per year, with one fatality every two years. Keep this in perspective now. This is less than the death count from cattle, which gore or stomp 20 Americans to death each year. Remember that five people die every hour in accidents involving cars.

How about our risks here in California? As a state we seem to have kept pretty good records. In the 85-year period from 1926 to 2010, there have been 110 confirmed unprovoked shark attacks along our coastline, ten of these were fatal. Ten deaths due to shark attacks in 85 years is about the same risk of dying from a tsunami in California.

The risks are very, very low, but for those whose minds drift towards sharks when they enter the water, two counties have experienced a little over one-third of those 100 attacks, San Diego with 17, Humboldt with 15, and Marin, Monterey, and Santa Barbara counties have each recorded 10 attacks over the past 85 years. Sonoma and San Luis Obispo counties have each experienced eight. Santa Cruz County is next with 7 reported shark attacks in 85 years and none of these was fatal.

While less than one person dies every year in U.S. waters from a shark attack, people around the world kill an estimated 100 million sharks each year, many for simply their fin, which is cut off while the shark is tossed back into the ocean to die. Doesn't seem quite fair, does it? For those people who are brave enough to go to the beach and actually enter the water, the chance of being attacked by a shark is about 1 in 11.5 million, and a person's chance of getting killed by a shark is less than 1 in 264 million. Meanwhile over 4000 people drown in the U.S. every year.

Article 9

The Biggest Animals That Ever Lived

Hadrosaur footprints, Salema, Portugal.
(© 2011 D Shrestha Ross)

I think we all have a fascination with extremes, which is why the Guinness Book of Records provides so much entertainment. When it comes to the animal kingdom, most of us probably think of dinosaurs as the biggest creatures that ever inhabited the Earth. And some of them were truly massive.

While exploring along the south coast of Portugal in April 2011, we had read about the existence of some Cretaceous Age (145 to 65 million years old) dinosaur tracks exposed in rocks along the coast somewhere near the small town of Salema. I decided I was going to find them. So we headed towards Salema, about 4 hours south of Lisbon on the Algarve coast, found a place to stay overlooking the Atlantic Ocean and headed down the hill to the beach. Beautiful rocks exposed in the coastal cliffs, waves gently lapping on the sand, and an incredible blue sky surrounded us. Even if I didn't find the dinosaur footprints, it would have been worth the trip.

Hiking along the beach, I climbed up onto every ledge, and scoured every slab of rock I could reach. After about an hour of searching, there they were - a trail of ten perfectly preserved, very large footprints embedded in a sandstone slab about ten feet above the beach. There were no signs or any indication that they were there, and no one seemed to know anything about them; just a short vague reference in one of the two Portugal guidebooks we had with us.

As a geologist, I have to say this was pretty exciting. You just don't find Cretaceous dinosaur footprints in your backyard every day. After showing the photographs to two paleontologist colleagues back at the University, they both told me they were most likely hadrosaur tracks. Hadrosaurs are one of the large ornithipod (bird-hipped) dinosaurs that evolved in central Asia about 100 million years ago, spread west across Europe and also migrated east across a land bridge and into North America.

These were duck-billed, vegetarian dinosaurs, and some species were up to 40 feet long. Even at five tons they could walk upright on their large muscular hind legs. They had three toes and the footprints I found were about 16 inches across. These were big animals, but there have been much larger dinosaurs discovered.

Reconstructing a dinosaur's length and weight is difficult, especially when all you have is a few bones. But there have now been nearly complete skeletons of most species unearthed so that we have some degree of confidence that we are getting a reasonably accurate picture of their size. There is some agreement that the largest dinosaur that has been discovered to date is the Argentinosaurus, named after the country where the bones were recovered. It was also a plant eater, and is believed to have been about 120 feet from head to tail, and the best estimate is that it would have weighed about 100 tons - the equivalent of 40 Hummers or 20 average African elephants. Pretty impressive.

So what do we have around today that would even come close to the size of Argentinosaurus? You might be surprised to know that the largest animal that ever lived, at least that we have discovered so far, in terms of weight, is still alive in the ocean today - the blue whale. A full-grown blue whale can weigh up to 200 tons, nearly twice the weight of the largest dinosaur! A blue whale can reach a maximum length of 110 feet, nearly as long as the Argentinosaurus. And perhaps surprisingly, the female is typically larger than the male.

Everything about a blue whale can be expressed in superlatives. Their tongues alone can weigh as much as an elephant; their hearts, as much as a small automobile. And they reach this massive size on a diet composed almost exclusively of tiny shrimp-like animals, known as krill. When feeding actively, an adult blue whale can consume 3-4 tons of krill a day. Blue whale calves are huge when born, being about 25 feet long and weighing as much as 3 tons. They feed completely on their mother's milk for their first year and gain an impressive 200 pounds a day.

Blue whales are also among the Earth's longest living animals. Biologists have discovered that by counting the layers of a deceased whale's wax-like earplugs, they can get a close estimate of the animal's age. Much like humans, the oldest was believed to be about 110 years old, although more typically they will live to be 80 to 90.

Whalers in the 1900s, seeking whale oil, nearly drove the blue whale to extinction. Between 1900 and the mid-1960s, about 360,000 were slaughtered. The establishment of the International Whaling Commission in 1966 finally provided protection, although recovery has been modest. There are now still only about 10,000 to 25,000 blue whales swimming throughout the world's oceans.

For reasons still unknown, a large blue whale washed ashore near Pescadero on September 6, 1979. After several days of jurisdictional uncertainty, biologists and students from UC Santa Cruz began the long and unpleasant task of flensing, or removing the blubber and flesh from the whale. The task took about a month, and then with a helicopter and a large truck, the remaining bones were transported to Long Marine Laboratory. It took about a year for natural processes to clean away the remaining flesh and oil.

At 87 feet long, the blue whale skeleton standing beside the Seymour Marine Discovery Center today is believed to be the largest on display anywhere in the world. Not only is the blue whale the biggest animal that ever lived on Earth, we have the largest displayed skeleton in our own backyard.

Blue Whale skeleton, up close.
(photo: Rachel Leonard, Seymour Marine Discovery Center)

Part VII: Marine Life

ARTICLE 10

The Life of a Blue Whale

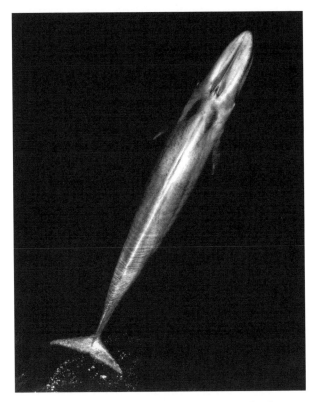

Blue Whale, the biggest animal that ever lived.
(photo: T Bjornstad, NOAA Fisheries via Wikipedia Commons)

Standing beside Ms. Blue, as she is affectionately known by the staff and volunteers of the Seymour Center, one can't help but feel dwarfed by the skeleton of this impressive creature. As far as we know, she is one of only four blue whale skeletons displayed in all of North America and one of only eight in the entire world.

- Santa Barbara Natural History Museum has a 72-foot specimen
- New Bedford Whaling Museum mounted a 66-foot specimen
- North Carolina State Museum displays a 65-foot specimen
- Worldwide there are blue whale skeletons on display in Canada, South Africa, New Zealand and Australia.
- At 87 feet long, Ms. Blue is the largest displayed whale skeleton in the world.

Dave Casper, our marine lab and campus veterinarian, took on the responsibility of reconstructing Ms. Blue when she was relocated next to the Seymour Center about 13 years ago. When the project was completed and dedicated in February 2001, Dave gave a very touching talk about what her life must have been like. The following is taken from Dave's words on that special February day 13 years ago.

The best estimate was that Ms. Blue was about 50 years old when she died in 1979, and at 87 feet long, was mature but still growing. She would likely have been born in the Sea of Cortez between December and March. Under ideal conditions she could have lived another 40 or 50 years and may have grown as long as 90 or 100 feet.

A good guess is that she was born about 1929, at the beginning of The Great Depression. This period, it seems, was also a depressing time for blue whales. In 1930, during her first year of life, 30,000 blue whales were killed in the Antarctic Ocean in a single season. We believe that there were probably about 200,000 in the southern Polar seas at that time. But before blue whales were protected in 1966, whalers had slaughtered 350,000 of the world's largest animals.

The North Pacific blue whale population was always considerably smaller, perhaps 6,000 in pre-whaling days. Norwegians sent factory ships to the Baja California in the 1920s, however, and took a total of 1,300 whales, over 20% of the population. During her lifetime, about 9,500 blue whales were killed in the North

Pacific, and we believe that she was one of only about 2,000 that survived the hunting.

Somewhat surprising for a state as environmentally conscious as California, one of the last two whaling stations in the United States was located in San Francisco Bay at Richmond. They were actively killing and processing an average of 175 finbacks, humpbacks and sperm whales a year up until 1971.

Our whale was about 25 feet long and weighed about three tons at birth. She would have doubled her weight in the first weeks of life and at six months old, would have been about 50 feet long and weighed in at perhaps 23 tons. After a long spring migration north she would have been weaned somewhere in the North Pacific in June or July. Blue whales typically spend the summer months in the North Pacific feeding and then may head south in the winter to the tropical waters off Costa Rica and Ecuador.

By the time she was ten, Ms. Blue would have been having a calf of her own every two or three years. In September when she died she would have been starting her fall migration back towards Baja. She was not pregnant so had likely weaned a calf that summer. By July of 1979, at the end of her nursing period, she would have lost nearly 50 tons in nurturing a 23-ton calf. She would have been eating three to four tons of krill a day during the summer to make up for that weight loss.

Her death off the coast of California in summer was a harbinger of change for North Pacific Blue Whales. We now think that something was starting to happen offshore to their food supply, the krill. Blue whales have always migrated both south and north along the Baja coast to feed. However, up to the time she died, they tended to avoid the U.S. coast in summer and fed farther out at sea.

She died at the start of a ten-year period when blue whale sightings along the west coast had doubled. This increase is too large and happened in too short a time to be accounted for by reproduction alone. Climate shifts throughout the Pacific Basin affect productivity, including krill abundance, and can span several decades. These changes in the distribution of their food supply probably drive whales inshore to feed during certain years. Perhaps that's why she was close to shore when she died.

We don't know why she died. Her body was too decomposed. But she's home now, and has become the symbol of our marine lab, and also of the often fragile state of life in the sea.

ARTICLE 11

Fertility, Food Chains, and Fish

The waters off the coast of California, like Peru, are some of the most biologically productive on the planet, because of a process known as upwelling. During the spring and early summer months, winds from the northwest dominate along the California coast, and help drive the offshore California Current southward. The surface waters of the ocean, however, are also influenced by the Earth's rotation.

This process, known as the Coriolis effect, causes surface currents in the northern hemisphere to be deflected ninety degrees to the right of their direction of movement. As a result, the surface waters off California tend to move offshore in the spring and early summer, and are replaced by bottom water through upwelling. This deeper bottom water is typically rich in nutrients, such as nitrates and phosphates, from the decomposing organic matter that is constantly sinking to the seafloor from surface waters.

The combination of the nutrients, which serve as fertilizer, and the exposure to the longer days and sunlight of spring and summer, lead to enhanced photosynthesis or blooms of the phytoplankton, which are the small floating algae. These microscopic plants, such as diatoms, are in turn fed on by the zooplankton, or the small floating animals like krill. The growth of the small plants and animals serve as the base of the food chain that provides for all of those marine animals higher up the food chain, the fish, sea birds, and marine mammals.

Coastal upwelling also influences weather patterns. Along the northern and central California coast, upwelling lowers sea surface temperatures and increases the frequency of summer fog. The cold surface waters chill the overlying humid air so that saturation occurs and fog forms, just like the condensation of moisture that occurs on a glass when you bring an ice cold drink outside on a warm day.

Globally, upwelling regions only constitute about 0.1% of the total surface area of the oceans, but these regions account for an astonishing 95% of the global production of marine biomass, and over 20% of the world's fish catch. The major upwelling areas occur off the west coast of continents. In addition to California, these include the rich fishing grounds off Ecuador, Peru and Chile, and also off northwest Africa.

The fertile waters offshore California have been fished for as long as there have been humans occupying the coast. Native Americans stayed close to shore, fishing in the bays, estuaries and shallow near shore waters. The Russians, Chinese, Japanese, Italians, Azoreans, Portuguese, and others who came later, all discovered different marine resources they could harvest from coastal waters. At different times over the past 150 years these included abalone, albacore, anchovies, crabs, salmon, sardines, sea otters and sea urchins, shrimp, squid, rockfish, whales, and just about everything else that could be eaten or had any value to humans.

Many fisheries that flourished for years and supported entire industries and cultures have now been relegated to the history books. Each fishery or resource had its own history and reasons for decline, although in some cases it was overfishing, or simply taking out more of the resource than could be continuously sustained. In other cases it was economics or government protection (sea otters and whales).

As we have improved and expanded our ocean monitoring efforts, however, we have also discovered that there are major shifts in ocean climate that can have a significant influence on the abundance and distribution of marine life. In the Pacific these shifts or cycles have now been recognized as lasting several decades and are known as Pacific Decadal Oscillations or PDO cycles. The alternating dominance of sardine and anchovy populations in our coastal waters is a result of these changing offshore ocean conditions, water temperature and nutrients primarily, and has now been understood as a "regime shift".

Our coastal waters are highly productive, but the marine life out there may change from decade to decade, with the best example being the cyclical alternating abundance of sardines and anchovies. The year 1977 marked the end of a 30-year long, cooler Pacific Decadal Oscillation. That year the commercial catch of anchovies was about 110,000 tons, while only five tons of sardines were caught.

In 1978 there was a pronounced shift to a warm phase of the PDO, which has lasted about 27 years. We now seem to have returned to a cool phase again (this shift seems to have taken place about 2005). Looking at the commercial catch for California during the middle of this warm phase, from 1995 to 2005, about 45,000 tons of sardines were caught on average each year, and only about 7,000 tons of anchovies. These population shifts are believed to be due to the types and sizes of plankton that thrive under these warmer and cooler regimes, and that are favored by either sardines or anchovies.

Article 12

Crabs and Calamari

The fertile waters of Monterey Bay sustained many of the early inhabitants of the central coast. Each ethnic group seemed to find something different to harvest or exploit. The Japanese used hardhat diving to collect abalone at Point Lobos. Immigrants from the Azores, Portugal and even Norway went after the whales, and the Italians fished for sardines, anchovies and lots of other fish as well.

The fishing industry has changed like everything else along the coast over the past 150 years or so. Very few abalone remain, and there is no commercial abalone fishery. Whales have been protected since 1971. Many of the groundfish or bottom fish catches have gone downhill. Salmon have been in significant decline in recent years, due to both troubles in the watersheds and changing ocean climate. In response to dwindling numbers, the Coho salmon fishery terminated in 1993.

How big is the California commercial fishery catch each year? Well, the peak years are long gone, but reached 550,000 to 700,000 tons annually during the heyday of the sardine industry from about 1935 to 1945.

The offshore ocean conditions then shifted to a cooler phase of the Pacific Decadal Oscillation and the sardine population crashed, lowering the overall catch in a big way. The commercial fishery overall was depressed throughout the 1950s and 1960s, with total catch much lower than in earlier years, but relatively stable, ranging from about 150,000 to 200,000 tons of fish per year.

Beginning about 1970 and continuing to the present, however, fish catches improved somewhat, with fishers in most years landing

200,000 to 300,000 tons. While some fisheries declined, others have developed or expanded as new markets have opened up, many of these in Asia.

Before you read any further, take a guess at what California's biggest fishery is in tonnage. What do you think the top three or four fisheries are?

I have to admit that I was surprised. In 2012, California's biggest catch by far was market squid, coming in at over 108,000 tons, or 61% of the state's entire commercial catch! This is 5.5 pounds of squid for every person in California! The squid fishery has been gradually increasing for the past 30 years with the exception of large El Niño years, such as 1982-83 and 1997-98, when the catch goes to nearly zero.

When the fishery first developed in 1863, Chinese fishermen rowed small sampans offshore at night and used the light from torches or fires hung in baskets off the side of their boats to attract the spawning squid to the surface where they could be easily netted. They dried their catch at that time and sold it for export to Asia as a food staple and also a fertilizer. Italian fishermen who emigrated from Sicily at the turn of the 20th century introduced their own nets, which increased competition between the Italians and Chinese and led to the expansion of the market squid industry in Monterey Bay.

Today the commercial squid fishermen use boats with very bright lights, which can be seen from the shoreline, to attract these cephalopods. Sadly, these little guys live for only about a year and are terminal spawners, meaning when they spawn, they die. Life is quite brief for your typical squid. Over half of the California squid catch today is typically exported, with most going to China and Japan.

So what is number two on California's commercial catch list? Sardines! At 25,400 tons, sardines made up a little over 14% of the tonnage landed in 2012. Sardines and squid are by far the two biggest commercial fisheries, often trading off in the number

1 and 2 spots; but in 2012, these two fisheries totaled 75% of the entire catch.

How about number 3? Dungeness crab. Commercial fishers brought in nearly 13,000 tons of crab in 2012, which made up 7.3% of the total catch. Depending upon the year and the price, however, Dungeness crab is often the most valuable of California's fisheries.

Dungeness crabs typically have a two-year lifespan, and there are size limits and a fishing season so as to protect the breeding grounds and to insure reproduction. While there are striking year-to-year fluctuations in catch, the crab catch has been gradually climbing since the early 1900s when it was virtually non-existent.

Going down the list to number four is one I would not have guessed - the red sea urchin. The urchin population is declining overall, and most of the catch is exported. What is left helps feed the southern sea otter population. At 5700 tons, the urchin catch makes up about 3.2% of California's total commercial catch.

Somewhat surprisingly, at least to me, in 2012 the market squid, sardine, Dungeness crab and red urchin catch totaled 88% of California's total commercial landings! Two of these four are almost never found on menus in any seafood restaurant in California, however, and end up being exported overseas.

Article 13

Invasion from the Sea

A 66-foot long concrete and steel floating dock washed onto the Oregon coast near Agate Beach in early June 2012. The Japanese consulate in Portland confirmed that the dock was one of four used by commercial fishermen for unloading squid and other catch at the port of Misawa, that had been ripped away from the coast during the March 2011 tsunami. It took about 15 months for the floating structure to make the roughly 5000-mile trip across the north Pacific, traveling about 10 miles a day.

Scientists from Oregon State University's Hatfield Marine Science Center discovered that the dock contained an estimated 100 tons of encrusting organisms, or about 13 pounds per square foot. These included several species of barnacles, as well as mussels, sea stars, urchins, anemones, worms, limpets, snails and algae - dozens of species.

Although most of the individual species are unique to Asia, this smorgasbord of marine organisms is similar to what you might find on a wharf or piling along the coast of California.

The Oregon Department of Fish and Wildlife set to work scraping and bagging all of the organisms in order to minimize the potential spread of non-native species. But they also were clear in pointing out that there is no way of knowing if any of the hitchhiking organisms or their eggs or larvae had already jumped ship and headed for new homes along the west coast.

Invasive or introduced species are not a new issue or concern. In the San Francisco Bay-Delta, the problem dates back to at least the California Gold Rush, when barrels of eastern oysters were shipped west to San Francisco aboard transcontinental trains. Some of them, along with the eastern seaweeds they were packed in, found their way into bay waters and proliferated.

The problem got progressively worse over the years, however, as San Francisco Bay and its shipping channels and ports became a center for global shipping for the western United States. Today an estimated 7,000 cargo container ships and about 10,000 tankers call at ports in the Bay-Delta.

Each of these ships contains 10 to 12 million gallons of ballast water, pumped on board from some foreign port. The ballast water keeps the ships stable when they are empty but is discharged on reaching port. And that ballast water contains organisms and eggs from those distant waters.

As a result, San Francisco Bay is now widely recognized as the most invaded waterway in the world. More than 240 animal and plant species are reported to have taken up new homes and are thriving in the waters from the Farallons to Sacramento.

Asian clams, Chinese mitten crabs, Amur River clams, New Zealand carnivorous sea slugs, Black Sea jellyfish and Japanese gobies are just a few of the exotics that now inhabit the bay and its adjacent waters. In individual parts of the Bay-Delta complex, invasive species may make up 40-100% of the common species and up to 97% of the total number of organisms.

And the problem in San Francisco Bay isn't diminishing. Between 1851 and 1960, there was about one new introduced species each year. From 1961 to 1995, however, they came at rate of about one every three and a half months.

We continue to bring in more and more imports by sea, our cars, computers, clothes, and sports equipment, as well as 50% of our oil, some 8 million barrels each day, comes into US ports. Although research is underway to determine the effectiveness of treating ballast water so as to disinfect it before it's discharged, there is a long way to go before this is a common practice or required.

While the initial arrival of tsunami debris has brought the issue of invasive or introduced marine species into a more public light, we have been surrounded by introduced terrestrial plants for well over a century.

Some of these were intentionally brought to the Americas from Asia and Europe and form integral parts of our agricultural economy - citrus, avocados, olives, figs, and artichokes, to name a few. Then there are the others, so invasive and ubiquitous that many new residents are probably unaware that they were introduced from somewhere else: Scotch and French broom, acacia, eucalyptus, pampas grass, ice plant, poison hemlock and thistles, to name a few.

So while the Oregon Department of Fish and Wildlife were attempting to be very thorough in their systematic annihilation of every living organism on that floating dock, the cat may have been out of the bag for some time. Ninety-five percent of all of our imports to the US come by sea and whether the larvae or eggs are discharged in ballast water, or from a bit of tsunami debris, they are all invasive species, some of which present serious problems, all of which are very difficult to control.

Article 14

Aquatic Aliens

Chinese Mitten Crab.
(courtesy: New Zealand Ministry of Fisheries)

While the marine life growing on the section of dock carried from Japan to the Oregon coast recently brought attention to the issue of foreign aquatic species invading our shoreline, this wasn't a new event.

In California, introduced terrestrial plants have surrounded us for well over a century. Many of these were intentionally brought from Central America, Asia and Europe and today are integral parts of our state's agricultural economy - citrus, avocados, olives, figs and artichokes, to name just a few. Then there are the others, so ubiquitous that perhaps some residents are probably unaware that they were introduced from somewhere else: Scotch and French broom, acacia, eucalyptus, pampas grass, ice plant, poison hemlock and thistles, to name some examples from the Monterey Bay region.

There are a few foreign marine species that are actually being considered for introduction because of their perceived benefits. Asian oysters, for example, are better at filtering out water pollutants than native oysters. They also withstand disease better and grow faster than the natives. Biologists are currently considering introducing the oysters into Chesapeake Bay to help restore stocks and remove pollutants.

More commonly, however, invasive or exotic species become problematic because they typically grow fast, reproduce rapidly, have the ability to disperse widely, and displace or outcompete the native populations, whether flora or fauna. They may also prey or become parasitic on native species, can transmit diseases, or may even impact human health.

Some examples of the direct impacts of these invaders include clogging of navigable shipping channels and canals, damaging crops, and reducing commercial or sport fishing populations of fish and shellfish. Damage can be extensive with estimated losses and control costs for all invasive species in California, terrestrial and marine, being about $3 billion annually.

A few of the aquatic bad guys include Caulerpa, the New Zealand mudsnail, and the Chinese mitten crab.

In June of 2000, several divers went for what they thought was a routine swim through the eelgrass bed in Agua Hedionda Lagoon in northern San Diego County. Working on a restoration project, they were swimming transects, measuring the extent of the eelgrass bed and noting new shoots. Then one of the divers came face-to-face with a large patch of unusually green, beautiful feathery seaweed. This strange plant would later be identified as the first confirmed North American siting of what has been called "the killer algae", or Caulerpa, which has also invaded the Mediterranean.

While not a combative "killer" in the true sense of the word, this renegade aquarium plant grows rapidly and can form a smothering blanket over mud, sand, or rock, severely reducing native populations of seaweeds and seagrasses. This invasive weed can take over the natural eelgrass beds, which provide habitat for lobsters, flatfish and bass, threatening their populations. It also can spread or grow up to ten times as fast as the native seagrass.

To date it has been found in only this one lagoon in San Diego County but also in Huntington Harbor in Orange County. Eradication efforts are underway through a process of covering the Caulerpa with tarps and attempting to chlorinate the affected areas.

The New Zealand mudsnail is native to fresh waters of New Zealand and was first discovered in the Owens River of eastern California in 2000. While there has been little research done on its potential impact, it is believed that their populations can become very dense, thereby reducing populations of other invertebrates, which could have a significant impact on trout fisheries.

In 1998 in a small Palo Alto creek that flows into South San Francisco Bay, a 10-year girl old discovered a small creature that could have been out of a science fiction horror movie. It looked like a large tarantula but with hairy claws and long spiny legs protruding from a dark shell.

North of Sacramento near Rio Linda, a 13-year old boy found similar creatures about the same time in a drainage canal, and began catching them in buckets to show his friends. Two years earlier, in 1996, biologists found 45 of these exotic crabs trapped on fish screens at water pumps in the Sacramento delta. In 1998 they returned and found 25,000 caught on the screens in a single day.

The creature, the Chinese mitten crab, has turned thousands of miles of California waterways and canal into a bad movie. These crabs have shown up nearly everywhere in central California, from Alviso to San Francisco to Sacramento and as far east as Roseville in the Sierra foothills.

The mitten crab, named after the dense patches of hair on its claws that resemble mittens, is native to the coastal rivers and estuaries of China. It spread through Europe in the early part of the last century and apparently was first discovered in California in 1993 in San Francisco Bay.

It isn't clear whether these crabs came in ballast waters or were illegally introduced. The latter is believed highly likely, as there is a lucrative market for them in China where they sell for $10 to $20 apiece, prized because the crab ovaries are believed to provide magical powers after being consumed.

Imports and sales were banned in California by the Department of Fish and Game after it was learned that live crabs were being sold in both Los Angeles and San Francisco. The species can carry a lung fluke, and instead or bestowing magical powers, they can cause symptoms very much like tuberculosis that an infect anyone eating raw or incompletely cooked crabs.

The population has exploded and has presented California with some serious problems. These hairy crabs feed on a variety of bottom dwelling animals, from shrimp to young shad, and potentially eggs and juveniles of salmon and sturgeon. The crabs burrow into stream banks and levees, which can accelerate erosion and reduce levee stability and safety. They have repeatedly clogged water intake structures in large numbers.

Because of the environmental and economic impacts that invasive species are having on California's coastal and fresh water environments, a number of measures are being developed and employed in efforts to reduce the spread and contain the damage. Treating ballast water is one way in which these aquatic hitchhikers can potentially be controlled, but unfortunately, there is a long road ahead simply because of the large number of species that have already become established in the state's inland waters.

Article 15

Peregrines, Pelicans, and Pesticides

The pelican and the peregrine are each rather amazing and unusual coastal birds. We almost lost both of them several decades ago due to the side effects of a chemical that saved millions of lives. These birds both have interesting stories, and fortunately, due to the work of a number of scientists and conservationists, these two species are both still here to observe and appreciate.

But, I need to set the stage a bit here. Why did we almost lose these unique birds from a chemical that was saving millions of lives around the world?

We live on a planet where malaria is extremely widespread. It might not seem that way from the perspective of a wealthy country, where malaria is usually thought of as a problem that has mostly been eradicated.

Although progress definitely has been made, malaria is still endemic to over 100 nations, threatening millions of people. In 2010, there were an estimated 220 million cases of this disease and between 660,000 and 1.2 million people died, most of them children, the vast majority living in Africa. The actual number of deaths is not known with certainty as accurate data is unavailable in many rural areas, and many cases are simply not documented or reported.

Malaria today is a disease of the poor and easy to overlook. Only in the past few years has malaria captured the full attention of aid agencies and donors. The World Health Organization has made malaria reduction a high priority

The word malaria had its origin in the Italian word, mal'aria, or "bad air", because the disease was often associated with marshes and swampy areas and was thought due to the air in these damp locations. It wasn't until the late 1800s that scientific and medical studies discovered that the culprit wasn't the damp air at all, but a mosquito that carried a parasite, and which laid its eggs in stagnant water.

The parasite and the mosquito that transmits it have been around for a long time, in fact during our entire history as a species. Some Egyptian mummies even have signs of the disease. Malaria parasites have been found in mosquitos preserved in amber 30 million years old.

Throughout history, only a few civilizations have escaped malaria. Alexander the Great likely died of it, leading to the unraveling of the Greek Empire. Malaria may have also stopped the armies of both Attila the Hun and Genghis Khan. Malaria was so pervasive in ancient Rome that it was known as Roman fever. Those areas of Rome where the population was most susceptible to the disease were all marshy, swampy or irrigated.

At least three US presidents suffered from it, George Washington, Abraham Lincoln and Ulysses S. Grant. In the late 1800s, malaria was so bad in Washington, D.C., sometimes known as foggy bottom, that one well known doctor proposed erecting a gigantic wire net around the entire city.

By one estimate, nearly two-thirds of all of the casualties in the U.S. Civil War were due to disease and an unknown number of these have been attributed to malaria. During World War II, casualties from the disease in the Pacific exceeded those from combat. Some scientists believe that as many as one-half of the humans who have ever lived on the planet have died of malaria. This is a very serious disease.

Dichlorodiphenyltrichloroethane, or DDT, was the miracle chemical that was going to eradicate the mosquito and rid us of malaria, but it also had some unanticipated negative impacts.

Article 16
Pesticides and Peregrines
Making the Connection

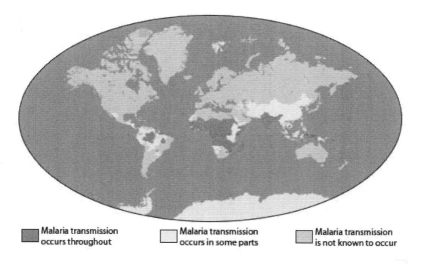

Occurrence of malaria around the world.
(courtesy: Centers for Disease Control)

DDT, or dichloro-diphenyl-trichloro-ethane, is the best known of several pesticides first introduced in the 1940s and 1950s. It was used widely during World War II to control the spread of typhus in Europe, and also was very successful in the South Pacific where it was used for malaria control.

The chemical was also helpful in the final eradication of malaria in both the United States and Europe, although by that time the mosquito and its breeding grounds had been nearly eliminated by other methods. Because of its properties as an insecticide, farmers quickly adopted it after the war as a general-purpose agricultural spray.

The World Health Organization in 1955 initiated a program to rid the world of malaria, primarily through the application of DDT. Overall this program was highly successful, particularly in areas where there was a higher standard of living, well-established healthcare programs, and less seasonal malaria transmission. These areas included a large portion of the South Pacific, the northern portion of Australia, most of the Caribbean, Taiwan, parts of northern Africa and much of Sri Lanka and India.

However, in other areas, primarily sub-Saharan Africa, because of the lack of a good healthcare system, the continuous life-cycle of the mosquito and the development of its resistance to the pesticide, application of DDT was very limited, and other methods of control were eventually implemented.

The connection between the increased usage of DDT as an agricultural insecticide and the impacts on wildlife, particularly birds, was a detective story that took some time and a number of different investigators to resolve. The effects on birds were simultaneously noticed in both Great Britain and the United States by a handful of scientists.

In 1962, the publication of the book, Silent Spring, by biologist Rachel Carson, is believed by many to have been the birth of the environmental movement in the United States. The book laid out the accumulating evidence on the environmental impacts of the spraying of large amounts of pesticides, and not surprisingly, was met with fierce criticism, particularly from the chemical industry.

Rachel Carson suggested that the widespread use of pesticides, including DDT, was a threat to wildlife, especially birds, and could cause cancer. She herself, was diagnosed with cancer in 1960, and died just two years after the book was published at the very young age of 56.

DDT had also been manufactured in Great Britain during World War II, and then used widely after the war against a variety of agricultural pests. One of its perceived benefits was its low toxicity to warm blooded animals, including humans, although there were indications of problems for both fish and, oddly, domestic cats.

It was soon discovered that treating seeds with DDT or similar chemicals, especially cereal grains, before planting, was an effective way to protect the young plants from insect pests. But this also created potential problems for large numbers of seed eating birds. And it was the connection that was discovered between high bird mortality around newly planted fields in Britain, and a declining number of peregrine falcons, which fed on birds, that led some concerned scientists to begin looking carefully at all the evidence.

This research was taking place in the early 1960s, at the same time Rachel Carson was looking into the same questions. Careful field observations and chemical analysis in Britain revealed either no eggs in peregrine nests, broken eggs, or very soft shelled eggs that didn't hatch successfully, as well as elevated levels of the breakdown products of DDT in the eggs.

As meticulous research continued, step by step, it was soon discovered that the pesticides were being biologically magnified moving up the food chain, from low levels in the soil, seeds or water, to higher levels in the peregrine's food supply, and ultimately to an accumulation in the falcon's fat tissue.

During breeding, the accumulated pesticide breakdown products led to a reduction of calcium and thinner eggshells, leading to the failure of peregrine eggs to successfully hatch. The population ultimately crashed and the peregrine falcon soon became an endangered species, in Britain and the US.

Article 17

Bringing Back the Peregrines from the Edge of Extinction

Glenn Stewart cleans off Sophie's beak.
(© 2011 Glenn Nevill)

In California and elsewhere, the peregrine falcon and the brown pelican both suffered serious reproductive failure because of the use of pesticides, particularly DDT, during the 1950s, 60s, and 70s. The biological magnification or accumulation of the pesticides up the food chain and their accumulation in the birds' fat reduced the amount of calcium in their eggshells. With thinner, or in some cases virtually no shells, fewer eggs survived to hatch.

Rachel Carson's book, Silent Spring, was a best seller, and generated widespread public concern about the use of pesticides.

President Kennedy requested his Science Advisory Committee to investigate the pesticide issues and, in essence, the group concluded that Carson's claims were valid. They recommended phasing out the use of persistent toxic pesticides. This did not happen right away, however.

The Environmental Defense Fund was established at this time by a group of scientists and attorneys whose primary goal was the elimination of DDT. A series of lawsuits were filed as research continued to reveal the connections between high levels of DDE (the breakdown product of DDT) and thin eggshells, in peregrines, pelicans and also the California condor.

Several years of court battles ensued, which involved the Environmental Protection Agency, the DDT manufacturers, and the Environmental Defense Fund. In 1973, a court ruling led to the banning of use of DDT in the United States, although with an exemption for public health purposes. Until 1985, however, the chemical was still being produced in the U.S. but exported to other countries.

Globally, nearly all use of DDT as an agricultural pesticide has now been eliminated, with the chemical largely replaced by less persistent insecticides. An agreement reached by over 170 countries in Stockholm, which took effect in 2004, outlawed use of some of the most persistent pesticides and restricted DDT for the control of malaria, where in many countries there are few affordable or effective alternatives.

The realization in the early 1970s that the peregrine falcon was nearly extinct in California led a local veterinarian, Jim Roush, to initiate an effort to try to help the population recover. A captive breeding program for peregrines undertaken by Dr. Tom Cade at Cornell University had been successful in the east and provided encouragement for an experimental program in California.

Jim got hold of Ken Norris, a Professor of Natural History at UC Santa Cruz (and one of the early leaders of the efforts to develop Long Marine Laboratory) and proposed a recovery plan. They had their share of challenges in obtaining funding and getting a program underway, but were able to find a dedicated young scientist and falconer in Brian Walton, who took charge. In 1975, the Santa Cruz Predatory Bird Research Group was formally established as part of the University of California Santa Cruz.

Much of their initial efforts were focused on locating the few remaining peregrine nests in California, not an easy task, in large part because the birds usually nest in high, remote, precarious locations, at least precarious for young birds and humans. Steep and nearly inaccessible coastal cliffs, bridges, offshore oilrigs and skyscrapers, are all recognized peregrine nesting sites. Flying around in helicopters and hiking through remote terrain was the approach used by Brian and his dedicated field crew.

As nests were located and cliffs and bridges were scaled, soft-shelled eggs were carefully removed and replaced with dummy eggs. Private funds were raised to pay staff and construct a facility in the lower quarry at UC Santa Cruz where these soft-shelled eggs were carefully incubated until the young peregrines hatched.

The baby birds were fed and cared for until they were large enough and then they were returned to their original nests, with the parents apparently completely accepting of having had eggs replaced with active teenagers almost ready to fly.

Through the efforts of the Predatory Bird Research Group (PBRG) and their supporters, the release of the peregrines hatched in captivity from the late 1970s through the mid-1990s was very successful. In fact it was so successful that the peregrine falcon was taken off the federal endangered species list. A statewide census by the PBRG in 2006 documented over 200 territories now occupied by healthy peregrines and there are now 250 pairs in California. This was a remarkable success story, which took place at UC Santa Cruz, thanks to a dedicated group of individuals and supporters.

Article 18

Offshore Feeding Frenzies

Pelicans, sea lions, whales, and dolphins were in an unusual feeding frenzy, which astounded all of us who watched from the bluffs, beach or water in late fall of 2013. From Natural Bridges to Capitola and beyond, and offshore as well, the food chain was in full bloom at a time when we don't normally experience this concentration of feeding animals.

It's really all about free food - somewhat akin to the sampling tables at Costco on a Saturday or Sunday afternoon. Only instead of cheese and crackers, offshore it's been all about anchovies.

While the waters off California are some of the most biologically productive on the planet, the timing of the 2013 feeding orgy was a bit abnormal. Usually the upwelling or overturning that brings the nutrients to the surface occurs in the spring and again in late summer or early fall.

The intense blooms of offshore life in October and nearly into December were quite unusual. And we don't usually see this many larger marine mammals or pelicans in these months. In addition to dozens of humpback whales, there was also a large pod of killer whales that was not wasting its energy chasing anchovies, but was focused on the sea lions – going for the steak rather than the pizza topping. It's life and death out there in ocean, and everybody has to eat.

Exactly why we saw these incredible numbers of anchovies in late 2013 isn't completely clear. There are some ideas that surfaced, perhaps the warmer water in the bay was more conducive to anchovy egg and larval survival right at that time, and the fish sensed that. But we aren't sure.

These tasty little silvery fish haven't been particularly abundant in recent years, and the anchovy fishery has been pretty modest in California. In 2012 the commercial catch was just 2,700 tons, or 1.6% of the total offshore fishery. But then, how many of you head off to the supermarket looking for anchovies? The anchovy catch picked up in 2013 and more than doubled to 6,620 tons and 3.6% of the total commercial catch, ranking number 5 for tonnage.

What were California's biggest fisheries last year? I often ask this question to different groups, and the answers I get are always surprising. Recently in my large oceanography class, I saw a hand rise ready to volunteer an answer: "abalone". Maybe my student thought this was a trick question, as abalone, are pretty scarce in California these days, almost an endangered species.

I also often hear "salmon". While there has at least been a commercial salmon season in the past several years after some closed years, in 2013 Chinook salmon was about 1% of the total catch in tonnage, but ranked number 3 in value. I think we tend to think in terms of what we eat most, which in California bears little resemblance to the commercial catch, in large part because so much of the catch is exported.

But before reading any further, take out a pencil and jot down what you think the four largest catches were in 2013. And a hint, one of them isn't anchovies, or abalone or salmon. I'll give you another hint, three of the top four are rarely seen in the fish section of your local market. While we spend a lot of time, effort and resources studying our offshore fisheries and trying to either sustain or rebuild them, three of our top four fisheries in tonnage are almost all exported.

So what are the big four? I have to admit that the first time I looked into this two years ago I was surprised. These shift around a little from year to year but there are 3 or 4 that are usually up there on the top of the list.

The overall 2013 commercial catch of all fish and shellfish was 181,685 tons, which had a value at the dock of $256 million, which averages out to about 70 cents a pound for the hard working commercial fishing community. Today the catch is only about 25-30% of what it was during the peak of the sardine industry in the 1930s and 40s.

As in prior years, the California market squid continued to dominate the commercial catch in 2013 and made up over 63% of the total tonnage. How many of you cook squid at home on a regular basis? While a few restaurants serve it, most gets exported. The total value at the dock for market squid over the past 10 years has ranged from almost $20 to nearly $73 million, which comes out to 24 to 34 cents per pound.

Squid fishing take place at night with bright lights attracting these little guys to the surface. This invertebrates have a very short life cycle, about a year, and the females lay lots and lots of eggs, so their population has been little affected by the volume of the catch each year, but more by oceanic conditions.

Number two is a more familiar dinner table treat, Dungeness crab, 15,584 tons and 8.6% of the catch. Crab moved up from number 3 in 2012 and ranked number 1 in value at $89 million.

Pacific mackerel came in at number 3 with nearly 9,000 tons, while number 4 was the Pacific sardine. The anchovy and red sea urchin were tied for number 5 at about 6500 tons each. Other than Dungeness crab, most of the other top five on the commercial catch list is exported to Asia.

So these six - squid, crab, mackerel, sardine, crab, anchovy and sea urchins - make up 88% of the total California catch. All the rest of the stuff you usually eat from the ocean, salmon, sole, shrimp, albacore, and the rest, are relatively small fisheries. Now if you were reading carefully, you would notice I changed subjects, diverted your interest, and never really answered the question of why there are all those anchovies out there. If I were absolutely certain I knew the answer, I would definitely share it with you.

PART VIII

THE GLOBAL OCEAN

Article 1

Heading Across the Pacific

A full moon over the shield shaped Mauna Loa volcano, Hawaii.
(© 2014 D Shrestha Ross)

Our ocean backyard is going to be a bit more global for the next four and a half months. We departed Friday, January 10, 2014 from Ensenada, Mexico on a 600-foot ship, the MV Explorer, headed for Hawaii. So this morning we will be about 200-miles further west as we head across the middle of the Pacific, hopefully with calm seas while we get acclimated to life on the water.

I'm teaching on the Semester-at-Sea program, administered by the University of Virginia, and which has about 550 students from 280 different colleges and universities enrolled in a worldwide voyage of discovery. Our Spring 2014 voyage will be the 50th anniversary of the program, which will probably be cause for some occasional celebration along the way.

Forty faculty members teach classes every day at sea, but while in port, educational, cultural, natural history and other field trips, travel and experiences will provide for a unique opportunity to study and learn about the global natural environment and also the diverse human communities.

Over the next four months the ship will have port stops in 16 cities in 12 different countries, including Japan, China, Hong Kong, Vietnam, Singapore, Myanmar, India, Mauritius, South Africa, Ghana, and Morocco. After having sailed about 25,000 miles across the Pacific, Indian and Atlantic Oceans, the ship will dock in Southampton, England on May 2nd. From my own past experiences on ships, I think it's fair to say that after this many months at sea, most everybody on board will be happy to be on dry land and a stable platform again.

The 2400-mile transit to the big Island of Hawaii takes six days, where one of my classes (Natural Disasters) will spend a long day exploring an active volcano, hiking across now cooled lava lakes, and climbing through lava tubes. The Hawaiian Islands are at one end of a 3000-mile long chain of volcanic islands and submerged seamounts that extend from Hawaii on the southeast to the Aleutian Trench to the north. The vast Pacific tectonic plate, which underlies virtually the entire Pacific Ocean, has been moving northwest over a hot spot or thermal plume in the mantle for millions of years. There have been about 100 active hot spots scattered across the Earth's surface over the past 10 million or so years, and active volcanism occurs where these molten plumes reach the Earth's surface. Yellowstone, the Galapagos, the Azores, Canary Islands, Samoa and Iceland are a few examples of hotpots.

As the Pacific Plate migrates over the Hawaiian hot spot, at a rate of only a few inches a year, the site of volcanic eruptions gradually shifts to the southeast, much like moving a piece of paper slowly over a burning candle. The big island of Hawaii consists of several very large shield-shaped volcanoes that haven't erupted for some time (Mauna Loa and Mauna Kea), and also the East Rift of Kilauea, on the southeast corner of the island, which is the site of active eruptions today. Offshore, 20 miles to the southeast, is the newest volcano, Loihi, which now rises about 11,500 feet from the seafloor. Based on current rates of activity, it should reach the surface in 30,000 to 100,000 years and will create another Hawaiian Island for more real estate development over time.

The northernmost seamount in what is known as the Hawaiian-Emperor Seamount Chain, the Detroit Seamount, was built up over the hot spot 81 million years ago, but has been slowly transported north. It lies adjacent to the Aleutian Trench and over time, it will be carried down into the trench and back into the mantle where it all started, recording the life, travels and death of a hot spot volcano.

The MV Explorer approaches the Port of San Diego.
(photo: Dale Frost, Port of San Diego)

ARTICLE 2

Crossing the Pacific Plate

The big island of Hawaii was a welcome port call after nearly a week at sea and I led 30 students in my Geologic Hazards class up to Kilauea, one of the most active volcanoes on Earth, for a day of volcanology. Many of the students were surprised that you could actually hike across a volcano and through an ancient lava tube.

Clouds of steam and sulfur gas are constantly being emitted from the caldera, and several hundred feet below the top of the crater where we stood was a molten pool of red-hot lava. While some enthusiastic students really wanted to get a close up glimpse of boiling lava, National Park safety regulations don't allow visitors to get that close to the edge, just in case someone decided to step too close or an eruption began unexpectedly.

We spent a half a day in Honolulu on January 18 for refueling, and then headed west-northwest for Japan. This was the longest leg of the entire voyage and one where we went days without seeing much of anything but water-perhaps a ship, maybe a lost bird or two, but not much else. The closest land at this point was directly beneath us, about 11,000 feet straight down. Somehow this wasn't a particularly comforting realization to those who had never been to sea before.

The Pacific Ocean covers about one-third of the entire surface of the Earth, and in fact, is larger than all of the land area combined. You can cross it in an airliner in half a day or so, which makes it seem a lot smaller. We moved considerably slower, however, at about the speed of a brisk bike ride. It takes us a whole lot longer to cross 6400 miles of water, in our case about 17 days. Our vessel could have gone a lot faster but there are two other considerations regarding speed: we needed to have enough days at sea to get the required lecture time in for classes, and fuel costs on a ship this size are very high, and as in a car, lower speeds mean better fuel efficiency.

Despite the immensity of the Pacific Ocean, and the ample opportunity for storms to create some uncomfortably large waves, we were fortunate not to have experienced any really nasty weather. The largest swells we encountered in the Pacific were only about 15 feet.

It doesn't take much movement to make some people uncomfortable, even on a 600-foot ship. But for the most part, people did well; at least if the number of students lined up at mealtime and the amount of food on their plates was any indication of wellness.

Even though the ocean out there seemed relatively calm, there were always swells or waves moving in different directions that crossed our path. As a result, the ship, despite its size, was constantly pitching and rolling, which tended to make walking around or standing up to lecture a constant balancing act.

Most large cruise ships claim to have stabilizers, as does the ship we traveled on. This tends to give passengers some false sense of comfort, even though these mysterious stabilizers are rarely if ever used because they increase drag and fuel consumption.

The color and surface appearance of the ocean as well as the cloud patterns were constantly changing throughout the day and from one day to the next. There were more shades of blue than I can describe and the water was quite clear, due in large part to a general lack of organic activity in the middle of the Pacific. There weren't a lot of nutrients or fertilizers 2000-3000 miles offshore to excite the phytoplankton or small floating plants, so there just wasn't a great abundance of visible oceanic life.

We also sailed well south of where large concentrations of plastic have been reported and saw virtually no debris in the water over our first 3000 or so miles. This was an encouraging sign and gives me hope for the ocean.

Being in the middle of 65 million square miles of Pacific Ocean, surrounded by vast volumes of seawater as far as we could see in every direction, made it difficult to comprehend how the human population could possibly have any impact on the sea.

While our impacts are most visible and obvious in coastal waters, the over seven billion of us that now occupy the planet are having measurable global ocean effects, whether it is changing the pH or acidity of the ocean, increasing its temperature, depleting many of the fish populations, or discarding plastic, we are the bad guys in this movie.

Article 3

Volcanoes in and Around the Pacific

View of Mount Fuji from the approach to the Port of Yokohama.
(© 2014 D Shrestha Ross)

Ten days in the middle of the ocean with no land in sight, in fact with no nothing but water in sight, is not something most of us normally experience. No matter that you are on a ship with 850 or so other people, there is still no solid land mass around to provide stability and comfort when you need it most.

Although images of cruise ships in tropical blue waters tend to lull us into a sense of complacency and calm, the wind and sea conditions can change quickly in the mid-Pacific and erase that image. While we started our spring 2014 Semester-at-Sea voyage (we try to remind students that this is a voyage of discovery and not a luxury cruise) in reasonably calm conditions, we had several days of rough seas and high winds, and even a 600-foot ship can move around enough to make you wish you were back on land.

After passing over the six-mile deep Japan Trench, the site of the magnitude 9.0 earthquake of March 11, 2011, we docked in Yokohama, Japan, on January 29, 2014. Japan lies between latitudes 30 and 45° north, comparable to the stretch along the west coast from Ensenada, Mexico to about Portland, Oregon. Its winter here though and it was cold, a surprising change from the tropical weather we briefly enjoyed in the Hawaiian Islands.

Entering Tokyo Bay and the harbor, Mount Fuji displayed a coating of snow and nearly perfectly symmetry and added a feeling of geological unrest to the horizon. Looking from Puget Sound, the massive cone of Mt. Rainier, framing the skyline of Seattle, presents an almost identical image. Rainer last came to life in the 1840s while Mt. Fuji has lain dormant since 1707. Both will almost certainly erupt again, however.

The perfect photograph taken at the ideal time of day, which rarely occurs, often shapes our images of so many famous natural landmarks and buildings. Fujiyama is no exception, and we were fortunate to have the entire mountain framed in the window of our cabin as we approached Japan. Minutes later the fog rolled in and it was gone for the remainder of our stay.

But the steep-sided, classic cone shape of Mt. Fuji was a striking contrast to the low shield shape of the Hawaiian volcanoes. In fact, nearly all of my students were surprised to see that Mauna Kea, Mauna Loa and Kilauea, the volcanoes that make up most of the big island of Hawaii, didn't look at all as they had imagined.

The Hawaiian volcanoes were formed of basaltic lava that came to the Earth's surface directly from the mantle and that is nearly all black or very dark grey in color. The volcanoes' low-relief shield shape stems from their magma's chemical composition, one very low in silica, which yields a very fluid lava that can flow long distances without forming steep slopes. While these lava flows can and do destroy homes and farm land, they move slow enough such that they rarely are a hazard to people. Where they meet the sea, the flows are also gradually expanding the land area of Hawaii. In time, lots will be sold as they have in the past and homes will be built.

The Cascade volcanoes, on the other hand, extending from Mt. Lassen in Northern California to southern British Columbia (and include the well-known peaks Mt. Shasta, Mt. Mazama or Crater Lake, Mt. Jefferson, Mt. Hood, Mt. Saint Helens and Mt. Rainier) all fit our ideal image of steep-sided, cone-shaped peaks.

These mountains, as well as the other volcanoes that form the Ring of Fire around the Pacific, were created from magma that passed through continental crust that was higher in silica on its way to the surface. This additional silica produces very sticky or viscous magma that doesn't allow the volatiles or gases to escape and as a result, commonly produces explosive eruptions and very steep sided cones. The sticky lava flows combined with fragments of rock and ash blown out under pressure solidify quickly to form the steep slopes and statovolcanoes.

ARTICLE 4

Perils of Foreign Ports

Karst topography along the Li Jiang, China.
(© 2014 Gary Griggs)

Shanghai has been described by my good friend, Sandy Lydon, as Las Vegas on steroids. If you arrive at night, the lighting on the dozens of architecturally interesting and increasingly taller skyscrapers definitely is over the top. Colored light shows, images in neon, constantly changing-advertising, it's all there like Las Vegas or Times Square; but its China.

Although we arrived at night, this wasn't the captain's plan. The ship was supposed to anchor at 8:00 am in order to maximize our time in port. But there were some unplanned surprises.

Virtually every port and harbor around the world has licensed pilots who know the navigation routes inside and out and who have responsibility for making sure each ship reaches the dock safely.

For Shanghai, there is a long route up the Yangtze River, and lots of shallow water where an unsuspecting captain could run a ship aground. The pilot meets the ship just offshore and then guides the captain and ship for miles and hours up the river channel to the port terminal. There are two pilot stations for Shanghai, however, and communication was confused so we waited for a pilot who never came. He was waiting at the other station.

By the time the miscommunication was discovered, the tide had gone out, leaving the channel too shallow for the MV Explorer to navigate. Coastlines around the Pacific Ocean have two high and two low tides each day, so we had to wait nearly 12 hours for the next high tide to raise the water level. You really can't rush the Earth in its daily rotation, nor change the position of the moon.

This did create havoc for all of the overland trips that were supposed to depart from the ship at 9:00 that morning, including one that I was scheduled to lead to Zhujiajiao, a water village of Shanghai with a history extending back over 1700 years.

But the sea waits for no person, and the tide had to go out and come back in again. What it did mean was that we arrived at 8:00 pm that night when the skyline was completely illuminated, rather than 8:00 am when the skyscrapers would have had much less personality.

I had a second trip to lead that did depart the following morning with 28 people bound for Guilin. This is the mountainous area depicted in so many Chinese paintings that looks like it came out of a Dr. Suess book, dozens of tall rounded peaks which the Li Jiang River passes through in dreamlike scenes.

This is typical karst terrain, which forms in limestone, and which was originally named after the karst region of former Yugoslavia. A large area of what is now southern China was covered with an ocean about 300 million years ago, and thousands of feet of calcium carbonate accumulated on the floor of that ancient ocean, primarily from the shells of gazillions of microscopic plankton, which populated the surface waters for millions of years.

Over the subsequent centuries of sediment compaction, continental collisions and uplift, the sediments were converted to limestone and elevated to form mountains. The stress of uplift created cracks and fractures in the rocks, and because limestone dissolves in acid, the slightly acidic groundwater flowing along these cracks began to create valleys at the land surface, and also dissolved caves in the subsurface.

As the valleys grew deeper and deeper over time, these bizarre Dr. Suess mountains stood higher and higher above the surrounding landscape. Today the Li Jiang River winds magically through these fairy tale mountains, creating a geologist's paradise. We took a 35-mile, six-hour boat trip down the river. While the temperatures hovered close to freezing all day, it wasn't enough to keep us from staying out on deck taking a lot of photographs. Every bend in the river brought a new set of peaks that seemed more interesting than those we had just passed. Fortunately, once you have your camera, digital photography is really cheap.

Article 5

Traffic at Sea

Do you ever wonder when you walk into Costco, The Gap, Radio Shack, the Apple Store, or just about any other retail shop, where all that stuff actually comes from? Its pretty safe to say that the great bulk of it comes from some place else and wasn't produced in the USA. Look at the labels on most of your clothes, your camera, computer, i-gadgets, you name it, and odds are that we probably didn't make it here.

It turns out, perhaps not surprisingly if you stop and think about it, that about 95% of all that stuff we import and then fill our houses, garages and storage lockers with comes into the United States by sea. Sailing into and out of a handful of Asian ports provided a whole new perspective on just how much stuff gets moved around the world's oceans by ships.

While I saw only a few vessels on our entire 6,500-mile voyage across the Pacific, once we hit the port cities of Asia, the number of ships-of all sizes and shapes-was overwhelming. Vessels are lined up in shipping channels and harbors like local commuters on Highway 1 at 5:00 pm: oil tankers, container ships, bulk carriers, and boxy car transporters full of Toyotas, Subarus, Hondas, and Nissans.

The oil tankers range from large to huge, sometimes referred to as VLCC for Very Large Crude Carriers, or even bigger, ULCC, for Ultra-Large Crude Carriers. Giving something an acronym seems to help diffuse its magnitude. You could put three soccer or football fields on the deck of one of these 1200-foot long behemoths. They don't usually play football on the deck but put as much as 3.5 million barrels of crude oil in the tanks.

We arrived in Singapore on February 23, 2014, and from the fleet of ships anchored offshore, the size of the container port and the number of cranes and containers waiting to be loaded, its clearly a huge shipping center. In fact, Singapore and Shanghai battle each other for bragging rights for being the world's busiest or biggest port.

There are different ways to measure port size, which complicates the contest, including total cargo tonnage, number of ships using the port, numbers of containers, and a few others. The ranking shifts back and forth from year to year between these two, but let's call it a tie. They are both massive, each moving over twice as much cargo annually as any U.S. port.

Seeing the number of ships coming in and out of these Asian magaports, and thinking about where so much of the stuff we import is manufactured, I guess it shouldn't be surprising that of the world's 20 busiest or biggest ports, 14 are in Asia, and 9 of these are in China. And my guess is that many of us have probably never heard of seven of these: Tianjin, Guangzhou, Quindao, Ningbo, Qinhuangdao, Dalian and Shenzhen. The factories in China are making huge amounts of stuff, and a lot of it is shipped across the Pacific to our retail outlets and waiting consumers.

The growth in these port cities is astounding by our standards, shoot, by any standards. I visited Hong Kong 30 years ago and looked north out over an imposing barbed wire barrier into China and what was then rice paddies and a rural village. That small village, Shenzhen, became a Special Economic Zone, is now the world's 15th largest port, and is home to 12 million people, all in less than 30 years! I don't think China has a word for Environmental Impact Report, however. For comparison, Los Angeles County is the most populated county in the entire United States, just hit 10 million people, and this took over 200 years.

The United States has two ports in the top twenty: South Louisiana and Houston. Our busy ports of Los Angeles, Long Beach, or Oakland don't even make it onto that list of the top 20.

Container ships, along with oil tankers, now really rule the seas. We have seen hundreds of these over the past month. The development of large containers that could be conveniently stacked revolutionized transport by sea and land. These giant metal Legos made loading and unloading cargo much faster and far more efficient.

Over time, in order to save costs, container ships like oil tankers, were built larger and larger in order to carry more and more containers without requiring many additional crew. On an average day, between five and six million containers, stacked on ships, are being moved around the high seas, stuffed with everything from plastic bath toys, to Nike running shoes, computers, microwave ovens, and everything in between.

Container ships are not immune from large waves, however, and with these metal boxes stacked 7 or 8 high above deck, a surprisingly large number of these go overboard during storms. Shipping companies estimate that somewhere between 2,000 and 10,000 containers annually end up in the ocean. This means 5 to 25 of these end up in the water every day, and represent a significant risk to other vessels for the several weeks to months that they can stay afloat. Robert Redford's sailboat hit one. Some appear to sink quickly, but all of them end up scattered across the sea floor beneath the shipping lanes of the world like sunken treasure chests.

Article 6

Containers at Sea

The MOL Comfort container ship breaks in half in the Indian Ocean, 2013. (courtesy: Kopkov Nikolai and LPG Gaschem Arctic)

Earlier this year, on a voyage from Rotterdam to Sri Lanka, the Danish registered container ship, Svendborg Maersk, encountered hurricane force winds and 30-foot waves off the coast of France. After arriving at the Spanish port of Malaga, the ship's crew checked their cargo and discovered that stacks of containers on deck had collapsed and over 500 were missing.

On any typical day, between 5 and 6 million containers are on the high seas, moving between ports and full of everything from potato chips to refrigerators. But not all of them make it to their destination, as the crew of the Svendborg Maersk found out. Theirs was the largest ever reported single loss.

As the ship was undergoing repairs in Spain, the company's vice president of operations, stated that they were examining their procedures "to avoid similar incidents in the future." They also had to contact all of their customers to tell each of them that their shipments were at the bottom of the Atlantic Ocean. Not the message you want to receive, ever. Only 13 of the over 500 containers were recovered.

Floating low in the water, these twenty to forty-foot long steel boxes present formidable dangers for smaller vessels. Most containers won't float for long, especially in heavy seas. Those that are refrigerated may be buoyed by their insulation, and the use of polystyrene as packaging can also keep them afloat longer. One marine insurer reports that a 20-foot container can float for up to two months, but a 40-foot container might float more than three times as long. Not the sort of surprise you want to encounter in a small vessel in the middle of the ocean.

Seven years ago, the container ship M/V Ital Florida left Hong Kong on April 20, for her maiden voyage, which was interrupted by what has become known as a "Stack Attack". At nearly 800 feet long and weighing over 36,000 tons, this wasn't a small vessel by any measure.

The new ship encountered 22 to 32 foot waves in the Arabian Sea between June 16 & 19, 2007, not enormous by rogue wave standards, but with hundreds of containers stacked 12 wide and 6 high on the deck, things began to unravel on the brand new vessel in a hurry.

The containers shifted under the wave impact and over 40 were lost overboard. Who knows what they contained, but they ended up under 10,000 of so feet of water on the pitch-black floor of the Arabian Sea.

Fortunately, before matters got worse, they were able to limp through the Red Sea and Suez Canal without encountering any pirates. They reached the port of Alexandria, Egypt, where the ship and load were straightened out.

The question that is always asked after these incidents is whether this was a result of bad weather, improper securing or lashing of the containers, overloading, or some combination. Failing to adequately secure the increasingly larger loads of containers against the stresses experienced during heavy seas appears to be the most frequent factor in failures.

Another issue, which appears to have been at fault in some even larger container ship disasters, is the loading and weights of the containers. On June 17, 2013, 200 miles off the coast of Yemen, the 5-year old, 90,000-ton MOL Comfort literally snapped in half. Both forward and aft sections of the Comfort, as well as the entire load of 4,500 containers, went to the ocean floor. One factor being investigated is whether uneven loading of containers contributed to excess stresses on the hull.

These ships carry huge loads and frequently sail close to their maximum permissible stress or bending moments. A critical factor in the stresses acting on a ship is the weight distribution of the load of hundreds of containers. One uncertainty in the stability of container stacks is that these giant steel Legos are frequently not weighed, or that some shippers frequently understate the weight of their containers in order to reduce freight charges.

Not knowing how much your cargo weighs and how the weight is distributed along the length of a container ship can introduce all sorts of problems regarding the stress a vessel must endure at sea. Think of a teeter-totter with 300-pound NFL tackles on either end.

A proposal was put to the International Maritime Organization several years ago that containers be weighed before they were loaded on boad. Many shipping groups objected to the proposal as expensive, time-consuming, and extended their port time. Perhaps not surprisingly, nothing was done.

Exactly what led to the MOL Comfort's sinking will probably never be known, as the answer lies deep beneath the waters of the Indian Ocean. As I write this we are sailing across the Indian Ocean from India to Mauritius, hoping for calm seas.

Article 7

The Birth of the Indian Ocean

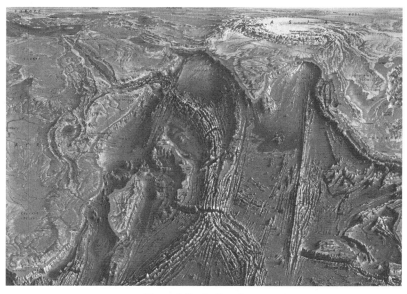

Floor of the Indian Ocean
(painted by Henriech Beeran, based on bathymetric studies by Bruce Heezen and Marie Tharp, Lamont-Doherty Geological Observatory)

The Indian Ocean was surprisingly calm during our 6-day crossing to the small island of Mauritius, where we stopped to refuel. I saw absolutely nothing around the ship on this entire transit but one bird and a few flying fish. The water is a brilliant deep blue, very clear, and a sure sign that there isn't a lot of life out here. There aren't many nutrients in the middle of the Indian Ocean, so there is nothing to sustain marine life.

While this 2,100-mile stretch of ocean appears homogeneous from a ship (read monotonous after 6 days of empty ocean), the seafloor beneath us presents a totally different picture - a complex jumble of ridges and fracture zones, seamounts, plateaus and rift valleys - which are all part of a complicated system of massive tectonic plates that has torn continents apart over the past 160 million years.

Africa, Antarctica, Australia, Madagascar and India were all joined together at that time in a huge supercontinent known as Gondwanaland. As a volcanic ridge opened up under this landmass, large fragments began splitting apart and the Indian Ocean was born. That ridge, or spreading center, appropriately named the Mid-Indian Ridge, now looks a bit like the stitching on a baseball as it traverses the sea floor.

In the north, this rift has split the Arabian Peninsula away from Africa by creating the Gulf of Aden and the Red Sea. The seam then bifurcates and heads southwest into Africa where the African Rift Valleys are gradually wrenching East Africa apart.

About 1,000 miles east of Madagascar, the Mid-Indian Ridge splits again, with one branch heading southwest, rounding the Cape of Good Hope and continuing into the Atlantic Ocean to become the Mid-Atlantic Ridge. Another branch heads southeast, passes between Antarctica and Australia and then veers north into the Pacific to become the East Pacific Rise.

This 40,000-mile long, global rift went essentially unrecognized until the 1950s when oceanographic institutions began to acquire surplus navy ships and head off on distant expeditions. As ocean bottom sonar records were gradually collected, the fragments of this global system of cracks in the ocean floor were pieced together and revealed the enormity of this feature.

The presence of active volcanoes along this ridge soon led to the concept of sea floor spreading, which provided a mechanism for how continents had drifted apart over geologic time. Hot molten magma from deep within the Earth's mantle surfaces along these fractures, cools to form ocean crust, and then spreads laterally away from the ridges like massive conveyor belts, carrying sea floor as well as continents.

And this is precisely what happened as the Indian Ocean opened up. One hundred and fifty million years ago, Madagascar and India broke away from Africa, along with Australia and Antarctica. Madagascar didn't move far, but India started on a 2000-mile trek, colliding with Asia 50 million years ago, pushing up the Himalayas, the highest mountains on Earth.

And beneath these empty waters, unseen by any mariner, are topographic features that provide the evidence and record of rifting seafloor, continental breakups and drifting landmasses.

Article 8

Stories from Ancient Rocks

View of Table Mountain, Capetown, South Africa.
(© 2014 D Shrestha Ross)

Traveling nearly around the world on a ship in early 2014 that averages about 15 miles an hour made me acutely aware of how big the oceans are - really big. Although the Earth's continents are huge, we could squeeze them all into the Pacific Ocean and have a little room left over.

I've also come to realize how long four months is when you are on a ship. Someone once said that the only difference between being on a ship and being in prison is that you can't drown in prison. There is a little truth to that, but fortunately we were able to stop in a number of different countries and get off the ship to travel overland.

While the oceans are huge, they have been created and destroyed and have changed size and shape repeatedly throughout Earth history. The evidence for that is scattered across the sea floors of the world and also preserved in the rocks exposed on land. While the air and water are constantly in motion and forget where they have been and what they have experienced, the rocks don't forget their histories.

Between about 250 and 200 million years ago, roughly the time of Jurassic Park, the Earth's landmasses were distributed quite differently than they are today. There was a large northern continent, known as Laurasia, which consisted of ancestral North America, Europe and Asia. Below the equator, South America, Africa, Australia, Antarctica and India, were combined in another very large southern landmass known as Gondwanaland.

In between the two was a large ocean called the Tethys Sea. The Mediterranean is a western remnant of that ancient ocean. This ocean, like all oceans, was the site of sediment deposition. Zillions of microscopic planktonic organisms flourished in the surface waters over millions of years. When they died, their calcium carbonate shells sank to the floor of the Tethys Sea. Over time and with the increased pressure and temperature of burial, those calcium carbonate muds that accumulated on the seafloor were turned to limestones. In some places, exposed to greater temperatures and pressures, they were metamorphosed into marbles.

As Africa pushed northward into Europe, those limestones and marbles were uplifted to become the Alps, including Michelangelo's Carrara marbles, as well as the limestone and marble of classical Greek temples and of the Dalmatian coast.

Farther to the east, the summit of Mt. Everest at 29,000 feet above sea level consists of marine limestone deposited in the ancient Tethys Sea. This limestone was pushed upward during India's collision with Asia about 50 million years ago. The limestone towers of Guilin in Southern China also resulted from this same collision process.

Going back even further in time, about 550 million years ago, South America and Africa were separated by a pre-Atlantic sea, the Andamastor Ocean. Sediments deposited in that ancient ocean were uplifted to form the sandstones exposed today on top of Table Mountain, the iconic symbol of Cape Town, South Africa, when those two early continents collided before separating again 200 million years ago. The rocks don't forget their histories.

Article 9

Finding Yourself at Sea

The Atlantic Ocean 0,0 buoy (where the Equator meets the Prime Meridian)
(© 2014 D Shrestha Ross)

Being around college students who have spent virtually their entire lives surrounded by electronic devices with amazing capabilities frequently gives me the overwhelming urge to explain to them that those things permanently attached to their palms weren't always around. This usually produces nothing but dull stares, but doesn't discourage me from seizing on the next opportunity to remind them again.

Knowing exactly where you are today is as simple as pushing a button on your I-phone, but this is a relatively recent convenience. Before setting out on this voyage, a NASA scientist sent me a GPS (Global Positioning System), along with a compact device that measures sunlight intensity as a measure of ozone in the atmosphere.

NASA is interested in the changes and condition of the Earth's atmosphere and ocean and they don't get a lot of readings out in the middle of the ocean.

About noon every day I step out onto the deck with the instrument and the GPS to get a series of readings of solar intensity at that particular geographic location. Within seconds after I turn on the GPS I get a screen that shows me the number of satellites that are within range - usually 4 to 6 of them - orbiting around up there somewhere. And so NASA will know exactly where the solar reading was taken, probably to within a few hundred feet.

Two other satellites have been sending down precise measurements on the elevation of global sea level for the past 20 years. While we have had tide gages or water level recorders attached to piers and bridges around the world measuring sea level and its changes for about 150 years, none of these records provide an overall global picture. The record from each of the gages is biased by the stability of land they are anchored on. Some coastlines are rising (northern California or Alaska, for example), and some are sinking (New Orleans or Venice), so they all provide slightly different values for sea-level rise.

Satellites give us a precise global picture, in fact very precise. Based on two decades of measurements from space, we now know that sea level is rising at about 3.2 millimeters yearly, and the rate looks to be increasing. But how do those satellites orbiting the Earth, hundreds of miles up, know sea level with the precision of tenths of millimeters?

I have to admit from my background this seems like an impossible task. Sort of like having a space shuttle find and hook up with the International Space Station. Yet to the geodesists of the world, who routinely design instruments that know exactly where they are within millimeters, it's what they do.

One of the few exciting moments in our 6-day, 2700-mile, transit from Capetown, South Africa to Ghana, was passing the precise location where the Equator and the Prime Meridian cross.

To my amazement there was an instrumented buoy right on the spot collecting oceanographic and atmospheric data. Using satellite navigation, the ship found the exact location, circled the buoy, sounded the horn for 20 seconds, and we continued on our course. I doubt if many young people on board appreciated how amazing that was.

Article 10

Iceland-The Making of an Island

The Mid-Atlantic Ridge that slowly splits apart Iceland.
(© 2011 D Shrestha Ross)

Back in January, the first port on our 4-month voyage was Hawaii, a volcanic island in the mid-Pacific. The final stop on our return home was Iceland, another island that also owes its existence to volcanism. This is where the similarities end, however. Iceland extends almost to the Arctic Circle. While Hawaii's temperatures in May are in the balmy 70s and 80s, most of Iceland was in the 30s and 40s.

Iceland is also almost exactly 10 times larger. So while you can drive around Hawaii in a long day, Iceland, if the roads are all open, takes closer to a week, at least if you stop occasionally to take in the geology. We took 7 days to explore this cold, stark, sparsely populated and very tranquil, but geologically active country. With a population of 322,000, Iceland has less than 1% of California's population.

Everywhere you look, the island's geologic origins surround you - volcanoes and lava flows; hot springs and geothermal steam; glaciers and icebergs. The island nation also has the world's 3rd largest ice cap, after Antarctica and Greenland. If you have absolutely no interest in the Earth and its geologic activity and history, however, you can probably skip a trip to Iceland.

There are 22 "active" volcanoes on the island's central plateau, meaning that they have erupted within the last few centuries. Over a 100 more are "inactive" or "dormant", meaning they might or might not erupt again. You just never know about volcanoes.

Over an 8-month period between 1783 and 1784, Lakagigar or Laki, poured out the largest lava flow witnessed on Earth in the last thousand years. More damaging than the lava, however, were the massive clouds of poisonous hydrofluoric acid and sulfur dioxide compounds that killed over 50% of Iceland's livestock, leading to a famine that led to the loss of about one forth of Iceland's entire human population.

The volcanic ash and poisonous gases from the Laki eruption and its aftermath obscured the sun around the world for weeks, and caused a drop in global temperatures. This led to crop failures across Europe and is believed to have caused droughts and famine in India that led to the death of perhaps six million people globally, making the eruption the deadliest in historic times.

Some of you may have been stranded temporarily at an airport in northern Europe in 2010 when another of Iceland's active volcanoes, the somewhat difficult to pronounce, Eyjafjallajokull, erupted. The ash cloud drifted eastward and shut down most of northern Europe's airports, grounding hundreds of planes for days. Volcanic ash and airplane engines don't mix well.

Why all this volcanic activity and havoc in the North Atlantic? Well, Iceland straddles the mid-Atlantic Ridge, a 10,000-mile long stretch of the larger globe circling oceanic ridge system. Molten magma rising from deep within the Earth and surfacing along the ridge crest has created this island over the past 20 million years. The North American and Eurasian plates have been pulled apart along the ridge and are having a tug of war, stretching Iceland apart in the process.

PART IX

OCEAN EXPLORATION

Article 1

Drilling into the Sea Floor

Nearly 50 years ago, an ocean engineer who thought way outside the box, proposed that we mount conventional oil well drilling equipment on a large ship and drill a hole into the deep-sea floor. Willard Bascom's goal was to drill through the Earth's crust, which is much thinner under the oceans than under the continents, and find out what the underlying mantle was composed of. We believed that there was an abrupt boundary between the cold, rigid and thin crust and the hot, viscous and thicker mantle, which had been named the Mohorovicic discontinuity, or Moho for short, in honor of its discoverer, Andrija Mohorovicic. While we had indirect evidence about the nature of the mantle from the earth's gravity, from earthquake waves, and from the composition of meteorites, we really didn't have any direct knowledge about this huge, 1800-mile thick portion of the Earth's interior.

Bascom convinced the National Science Foundation in Washington D.C. to support this bold experiment, named Project Moho after the drilling target, and in the early 1960's a large ship was outfitted with the necessary drilling equipment to test the concept. The drilling rig, named the Cuss I, was successful in boring the first two holes ever drilled in the deep sea floor, one off San Diego and a second in water 12,000 feet deep off of Baja California.

There was a lot of initial speculation that drilling from a floating vessel through 1,000's of feet of water into the deep sea floor simply wouldn't work. One major concern was how could a ship manage to remain stable enough to remain at sea at one spot and drill for days at a time. But by using a series of anchors and surface buoys connected with taut wires, combined with thrusters or essentially large outboard motors on each of the four corners of the ship, the drilling vessel was able to maintain its position in deep water and successfully drill into the ocean bottom. The drill bit penetrated as far as 1,000 feet into the seafloor in this pioneering effort, still a long way from the Moho, which was perhaps five miles deeper, but this project was a major breakthrough and proved that this could be done.

Forty-six years ago in 1968, the year I began teaching at UCSC, the National Science Foundation launched the Deep Sea Drilling Project with a large, uniquely constructed ocean-going drillship named the Glomar Challenger. This was the beginning of an innovative partnership between the federal government, university marine scientists, and the drilling industry, to carry out scientific drilling into the sediments and rock of the ocean floor. By extracting long cores of sediments and sedimentary rocks from the ocean floor, scientists for the first time had history books that would help them understand the evolution of the Earth and the oceans and how they have changed over time. Prior to this effort, our understanding of the seafloor and the historic record that was preserved in its sediments was based on short cores of sand and mud obtained by dropping a weighted piece of pipe into the ocean floor.

No one knew at that time that the oldest rocks in the ocean basins were only 200 million years old, relatively young geologically speaking, and that they only recorded a fraction of the Earth's 4.6 billion year history. But as the drilling program continued and the ship criss-crossed the oceans collecting cores, the discoveries from the sediments quickly revolutionized our thinking about the history of the Earth and its oceans.

Throughout the history of the oceans, the record of all that has transpired in the overlying seawater was preserved on the seafloor as sediment and the remains of marine life that settle on the bottom and contained in the sediments. We now have an entirely new group of scientists, known as paleoceanographers, whose careers are dedicated to spending months at sea, recovering long sediment cores from the ocean seafloor through ocean drilling, and then painstakingly analyzing these sediments for the information and millions of years of history they contain. What have they learned?

The drilling ship Joides Resolution
(courtesy: Ocean Drilling Program, Consortium for Ocean Leadership)

Article 2

Ocean Drilling-Asteroids, Mass Extinctions & the Mediterranean

What began as a far fetched idea 50 years ago, drilling a hole into the seafloor from a floating vessel, was successful and soon opened up an exciting new era of ocean exploration which continues today.

The Ocean Drilling Program, by obtaining long cores of sediment and ancient rock from the floors of the world's oceans, has been making discoveries that have challenged old ideas, and brought entirely new concepts to light. Drillships have evolved and become more sophisticated, enabling scientists to drill in greater water depths and progressively deeper into the seafloor. Hundreds of cores have been obtained over the past 46 years, in water over 4 miles deep and penetrating as deep as 6,000 feet into the ocean floor. Each voyage is normally 2 months long and typically involves dozens of scientists from universities and government agencies around the world. The United States has provided much of the scientific leadership over the years, and scientists at the University of California Santa Cruz have played major roles in organizing and leading the scientific drilling program.

What have we discovered as these voyages have continued to probe the deep ocean floor? Drilling in the Caribbean uncovered proof that an asteroid struck near the Yucatan peninsula 65 million years ago, and not only contributed to the extinction of 60 to 70% of all plant and animals species on earth, including 90% of all of the plankton in the ocean, but also is believed to have had an important role in the die out of the dinosaurs.

An asteroid the size of Manhattan Island, traveling at about 45,000 miles per hour when it entered the Earth's atmosphere, struck the southwestern edge of the Gulf of Mexico. The impact left a crater about 150 miles in diameter as it ejected particulate matter and gases into the atmosphere producing several months of global darkness, which has often been called "nuclear winter". The fossils preserved in the sea floor sediments suggest it may have taken 500,000 years for organisms to recover. We now also know that Chesapeake Bay was created by a similar asteroid impact about 35 million years ago.

When the drill ship first entered the Mediterranean Sea in 1970 and begin to bring sediment cores back onto the deck of the ship, scientists were astonished to find hundreds of feet of salt deposits beneath the sea floor. The minerals included halite (sodium chloride or table salt), gypsum and other salts that are commonly found when seawater completely evaporates. This presented some problems for oceanographers because the water where the cores were collected was about 6500 feet deep. How do you evaporate the entire Mediterranean Sea? Dating the seafloor sediments revealed that the salts formed about 5-6 million years ago. The chief scientists on board pieced together a controversial new history for the Mediterranean to explain these bizarre salt deposits.

In order to have salt crystallize or precipitate out of seawater, you need to evaporate the seawater such as happens today in tidal flats and tide pools. This required both cutting off inflow of seawater into the Mediterranean by raising the shallow sill at Gibraltar, creating a dam in effect, and also meant that the climate at the time must have been hot and dry enough to evaporate the water trapped in the basin. Over perhaps 1,000 years the water progressively evaporated, leaving brackish salt ponds and lakes where hundreds of feet of salt were precipitated out.

Several hundred thousand years later, however, either the Straits of Gibraltar eroded or subsided, or the level of the Atlantic Ocean rose, and water flooded catastrophically back into the

Mediterranean again. In a little more than a century, a flow equivalent to 1000 Niagara Falls poured back in and refilled the entire basin. While there are still some differences of opinion about exactly how this all happened, and how many times, scientific drilling recovered sediment cores that revealed this dramatic series of events that took place 5 to 6 million years ago.

Article 3

Ocean Drilling Confirms Continental Drift

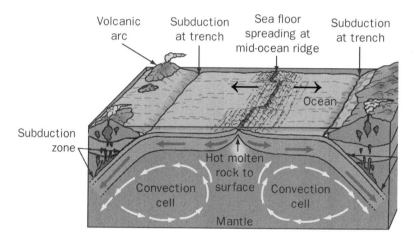

Convection cells surface at ocean ridges,
and drive plate motion.
(from the book, Introduction to California Beaches & Coast by Gary Griggs, 2010)

The theory of plate tectonics, or the concept that the surface of the Earth consists of a number of large rigid plates that move around relative to one another, generating earthquakes in the process, was proposed in 1968, the same year that the Deep-Sea Drilling Program launched its maiden voyage. Timing was ideal as the ability to sample the sediment and rocks of the seafloor provided a unique opportunity to test much of what the theory had proposed.

Plate tectonics evolved from the old idea of continental drift, and also a newer idea put forth in 1963 known as sea-floor spreading.

Continental drift dates back to the early 1900's, and grew initially from something that virtually all school kids appreciate today, that the coasts of South America and Africa fit together like a jigsaw puzzle. While other geological and fossil evidence accumulated over the 20th century supporting the idea that South America and Africa, as well as Antarctica, Australia and even India, were at one time joined as a large super continent named Pangea, scientists really didn't have a process or mechanism capable of moving entire continents around on the earth's surface. This was a bit troubling, and most geologists in the first half of the last century weren't at all ready to accept this idea.

Sea-floor spreading was put forth in 1963 as a mechanism to explain the drifting of the continents, and emerged at least in part, from the discovery of a 40,000-mile long undersea volcanic ridge system that looked much like the seam on a baseball. It was postulated that new ocean floor was created at the world-encircling ocean ridges, the Mid-Atlantic Ridge, being one example. As hot molten magma from deep within the earth rose towards the seafloor, it cooled and solidified, forming new oceanic crust. It was then carried laterally away from the ridge crest by a thick conveyor belt of crustal material. This conveyor belt was envisioned to operate much like the way an old window shade is pulled down or rolled out. One important difference was that you can pull down a window shade very quickly, but the seafloor was thought to be spreading at rates of only an inch or two per year.

If this idea was correct, then as the volcanic rock that formed the seafloor was carried further from the crest of the ocean ridge on this conveyor belt, it should be getting progressively older. In addition, the marine sediments that were slowly accumulating on the seafloor should be progressively older and thicker the farther you got away from the center of the Mid-Atlantic Ridge.

Something remarkable was discovered as the drilling ship began to bore into the ocean floor as it first crossed the Atlantic Ocean. The cores of sediment and rock that were brought back on board ship and analyzed, and then later dated by geologists, revealed that the sediments were very thin right at the crest of the Mid-Atlantic Ridge and the underlying volcanic rock was very young. Cores taken progressively farther from the ridge crest, as the ship headed for the coastline of North America or Europe, showed thicker and thicker deposits of marine sediments, with older and older fossils. When the core penetrated the volcanic rock beneath the sediments and this rock was radiometrically dated, it was also found to be progressively older the further they were from the ridge crest.

These deep cores confirmed the idea that the sea floor was in fact spreading laterally from the middle of the Atlantic Ocean and continents had been slowly drifting apart. The oldest rocks that have been recovered from the Atlantic Ocean floor are about 180 million years old, and have been recovered on opposite sides of the ocean, off the east coast of the United States and off the west coast of North Africa. By knowing the age of these volcanic rocks, and their distance from the ridge crest where they originally formed, its possible to determine how fast the sea floor has been spreading over the past 180 million years. This turns out to be about an inch per year or about the rate at which your fingernails grow. While this rate of ocean floor spreading seems very slow, over the past 180 million years it has been fast enough to completely open the Atlantic Ocean and separate and transport the continents on either side to their present locations.

Article 4
Discovering the Seymour Center

The Seymour Center's Blue Whale is the largest skeleton on display in the world.
(photo: Rachel Leonard, Seymour Marine Discovery Center)

Sitting on the coastal bluff on the lower west side of Santa Cruz, (and sometimes called the city's best kept secret), is the Seymour Marine Discovery Center. The Discovery Center has no desire to be a secret, however, and welcomes visitors of all ages from Tuesday through Sunday (and every day during the summer). The Center celebrated its 14th anniversary in 2014, but somewhat surprisingly, there are still occasional local residents I meet who either don't know the Marine Discovery Center is there, or don't know that it is open to the public. So I want to clear this up - it's here, and it's here for you.

The mission of the Seymour Center is literally about our ocean backyard. It's not simply an aquarium, although there are many tanks full of interesting and colorful marine life, and it's not a museum. It really is a Marine Discovery Center, where any visitor can discover what ocean scientists or marine biologists do; why studying the ocean and its life is so fascinating and engaging; where around the world the University's marine scientists explore and study; and what they have found out about the sea and its life in their expeditions and research.

One goal in the displays and exhibits is to change the perception that young people often have that all scientists wear white lab coats and mix chemicals in test tubes. Not so for UC Santa Cruz ocean scientists. Some tag and track seals in the ice and cold of Antarctica, while others study coral reefs and fish in the tropical Pacific. Marine geologists go out to sea in ships for several months at a time to drill deep cores from the ocean floor in order to unravel the 200 million year history of the ocean basins and past climate change. Others collect and study the plankton, the tiny plants and animals that are at the base of the food chain, the grasslands or pastures of the sea, and which make all other life in the sea possible.

The exhibits are designed to be understood by all visitors, and also to be engaging and educational. Did you know that salmon have otoliths, or very tiny ear bones, that record the history of their life travels through fresh and salt water, one day at a time, like a tape recorder? What's a tape recorder you might ask? Or by placing very small radio-transmitters on elephant seals, we have discovered that they swim thousands of miles out into the north Pacific to feed. We can add time-depth recorders to these small instrument packages. The recorders tell us that these seals spend most of their marathon swims beneath the surface, diving to depths of up to several thousand feet and routinely staying down for 30 to 45 minutes, holding their breath the entire time.

The 10-year anniversary a few years back brought a new shark tank to the Marine Discovery Center, and you can actually get up close and touch a three-foot long swell shark. Twenty feet away in one of the small aquariums you can see the embryos of several small swell sharks, in their three-inch long egg cases, growing slowly and waiting patiently for nearly a year until they hatch out and enter the real world.

If you head outside you can stand beside a skeleton of the largest animal that ever lived, a massive blue whale. While we tend to think of dinosaurs as the biggest creatures, the blue whale wins the prize, and can grow to over 100 feet long and weigh up to 200 tons. The whale mounted next to the Marine Discovery Center washed up on the beach at Pescadero in 1978, and through a massive and smelly salvage, cleaning and reconstruction effort, we are fortunate to have her on display for visitors. Another little known fact: not only is this the largest animal that ever lived, to our knowledge, this is the largest preserved and displayed blue whale skeleton anywhere on Earth. And it's all here in our backyard.

There are about 300 volunteers and docents who go through an extensive training program, and then donate their time each month to lead tours, work with school groups, keep the aquariums clean, feed the animals, answer questions at the touch tank, and all of the other jobs that are required to keep the Seymour Marine Discovery Center interesting, engaging and a valuable learning experience. If you haven't been, you owe it to yourself and your family and friends to stop by and visit Santa Cruz' best kept secret. And if you have been, come back and see what new things there are to explore. Spread the word, its not supposed to be a secret.

Article 5

Venturing into the Arctic

Frdtjof Nansen, c. 1889.
(photo: Ludwik Szacinski, via flickr)

Royal Dutch Shell Company has now repeatedly experienced the challenges of exploring for oil in the Arctic environment, and the Department of Interior is re-evaluating the whole enterprise and reconsidering the hazards and risks of Arctic drilling.

Until the late 1800s, the dominant view was that the Arctic was a solid mass, perhaps land covered with ice, or ice, but solid nonetheless. There was an earlier Arctic ship disaster in 1881 that ended an expedition planned to overturn the long-held ideas about a solid

Arctic. In July 1879, the Jeanette, a private U.S. vessel belonging to the owner of the New York Herald, but on a naval expedition, left San Francisco for the Arctic Ocean, after a major retrofit that included strengthening her hull to withstand the Arctic icepack.

The bold mission of the Jeannette, under the leadership of Lieutenant Commander George Washington DeLong, a veteran Arctic explorer, was to be frozen into the pack ice and drift to the North Pole. They were indeed frozen into the ice in September 1879, above Siberia near Wrangel Island. They drifted northwest for the next 21 months, gradually getting closer to the pole. It became clear that the ice cover wasn't solid, that there was an ocean beneath it, and that the ice at the surface moved around over time.

On the night of June 12, 1881, however, disaster struck as the pressure of the ice finally began to crush the Jeannette. DeLong and his men unloaded provisions and equipment onto the ice pack. The ship began to sink through the ice the following morning.

The rest of their trip is one of those adventures some of us like to read about while sitting in front of a warm woodstove, but none of us would like to endure. The expedition now faced a long trek to the Siberian coast, with little hope even then of rescue.

There aren't many people or settlements along the Siberian coast today, and there were even fewer in 1881. The voyage of the Jeannette had sounded exciting to the crew, the science had been interesting, but now they faced trying to survive in a very harsh and unforgiving environment. They didn't have any of the technology or life saving gear that an Arctic expedition today might routinely have - cell phones, GPS navigation, Gore-Tex, polar fleece, and air rescue if things got really bad.

Things didn't go well at all for DeLong and the crew of 30 officers and men and 3 civilians. Starting with sledges but ultimately getting into three small boats, they attempted to reach the mainland. One boat capsized in a storm with all men lost. The other two reached the Lena Delta, but in different locations, and began a long hike across the marshy, half frozen delta hoping to find a native settlement. As conditions got progressively worse and the men weakened, DeLong sent his two strongest men ahead for help. They did find a settlement and survived but George Delong and his eleven remaining crew all died from starvation and exposure. George Melville, the Chief Engineer, and his party of 11 reached a native village and did survive to tell their story.

One of the most important scientific outcomes of this tragic and ill-fated expedition was that in June 1884, almost exactly three years later, the wreckage of the Jeannette was found on an ice floe near the southern tip of Greenland. This discovery led great credence to the hypothesis that the Arctic Ocean wasn't a huge solid ice mass, but that the ice was in constant motion, and had moved the ship from the Siberian Coast, across the Arctic above northern Canada all the way to southern Greenland.

At this point, Fridtjof Nansen, our second bold Arctic explorer enters the scene. Nansen was a Norwegian who had one of the most interesting and fullest lives imaginable, the kind of person you would like to spend a week with.

He was born in 1861, and went on to become an explorer, scientist, inventor, diplomat, statesman, humanitarian, and Nobel Peace Prize laureate. In time, Nansen was to learn of the Jeannette expedition and outcome and become convinced that he could mount a successful expedition to the North Pole using the same locked-in-the-ice approach. He was a man who believed in going forward with bold ideas, even in the face of opposition, but he had to work his way up to this expedition.

The rural Norwegian countryside where young Nansen grew up shaped his nature. In the short northern summers the main activities were swimming and fishing, while in the fall, hunting for game in the forests took over. Long winter months were devoted to skiing and Nansen began to practice on improvised skis when just two years old.

At the age of 10 he defied his parents and attempted his first ski jump, ending up buried to his waist in the snow headfirst. This experience didn't diminish his enthusiasm for skiing and this was an endeavor that he would continue throughout virtually his entire life.

At school, Nansen didn't show any particular strengths and his studies often took a back seat to sports, or to expeditions into the forest where he would live "like Robinson Crusoe" for weeks at a time. These early experiences produced a remarkable degree of self-reliance, and he also became an accomplished skier and skater.

At the age of 15 his mother died suddenly, which led his father to move with his two sons to the city of Christiana. But young Nansen's athletic skills continued to thrive. At 18 he broke the world one-mile skating record, and in the following year, won the national cross-country skiing championship- an accomplishment he would repeat 11 more times.

In 1880, he passed his university entrance examination and decided to study zoology because it seemed to promise "a life out in the open air". What he later described as "... the first fatal step that led me astray from the quiet life of science" was to follow the advice of a professor and take a sea voyage to study Arctic zoology first hand. This expedition lasted 5 months and set the path that Nansen was to follow for many more years. But that story will have to wait.

Article 6

Nansen's First Arctic Adventure

We left Fridtjof Nansen at sea in 1882 in the North Atlantic. His first extended ocean voyage, which roamed the icy waters between Greenland and Spitsbergen, lasted 5 months and gave him the opportunity to make some important scientific observations. He was able to show, contrary to previous assumptions, that sea ice forms at the surface of the ocean rather than below, and also that the Gulf Stream actually flows beneath a cold layer of surface water.

The ship he was on, the Viking, was trapped in the ice close to an unexplored part of the coast of Greenland. Observing the coast while confined to the ice gave Nansen the idea of exploring the Greenland icecap and possibly even crossing it. While such an adventure may not seem so bold today, we have to think back to 1882 when we knew relatively little about this part of the world. Equipment, gear and navigation equipment were pretty primitive. You could easily die on such a trip.

Upon returning to Norway, Nansen was recommended as a curator in zoology at the Bergen Museum, where he was to spend the next six years of his life. But deep down, Nansen was an explorer. His mind kept returning to the Greenland icecap and the idea of traversing the vast ice-covered island, which had never been done.

Two earlier expeditions had set out from a settlement on the west coast, heading eastward across Greenland. Because there were no settlements on the more dangerous east coast, and therefore difficult for a ship to meet them, Nansen reasoned that those earlier treks would have had to return to the west coast anyway.

So he proposed going the opposite direction - starting from the more isolated east coast - assuming that he could find a safe location where his expedition could be dropped off by a ship, and then making a one-way journey towards the more populated west coast. His team would have no way to retreat to a safe base, and could only go forward - which he found to be the perfect challenge. The boat he was later to use in an attempt to reach the North Pole, the Fram, means forward in Norwegian, which was one of his guiding principles.

This became his exploration strategy. His thinking, planning and preparation for this first expedition was to foretell how he would undertake his next bold foray into the Arctic. But he had to get across Greenland first.

Nansen rejected the idea of a complicated plan with lots of people, supplies and equipment, which had been the approach used previously. Instead he hand picked five experienced adventurers, all adept snow travelers who were skilled in cross-country skiing, as was Nansen himself. They designed most of their own gear, sleeping bags, cooking stoves, and clothing, and then built lightweight sledges that they would tow themselves.

Nansen had great difficulty obtaining financial support because the Norwegian parliament and many others thought it was simply too risky and had little chance of succeeding. He ultimately did receive help from a Danish businessman and through a fundraising effort organized by the students at the university. Just before the expedition departed, Nansen sat for his Ph.D. defense, but he left without even knowing the outcome of the examination. His priorities were clear at this point and he was dedicated to his first major expedition.

In early June 1888, Nansen and his team sailed from Iceland to the eastern edge of Greenland on a sealing vessel, the Jason. Things from that point on didn't go smoothly or exactly as planned, but Nansen and his crew took it in stride. Thick pack ice kept them well away from the coast, but after several weeks of trying to find a route to shore, they departed in several small boats to row into Sermilik Fjord, which was their planned starting point for their trek across Greenland.

Nasty weather and sea conditions for the next two weeks, not totally unexpected around Greenland, kept them offshore and also swept them about 240 miles south of their starting point. For much of this this time they were camping on the ice as they felt launching the boats again would be too risky. The entire expedition was clearly dangerous, but this was deemed too dangerous.

When they finally reached the coast on July 29, Nansen felt they were way too far south, so he ordered his crew back into the boats to begin rowing north. He was a clearly a bold, confident and inspirational leader, and was able to keep his men motivated. For 12 days they toughed it out, battling the weather and navigating around ice bergs. On August 11, 1888, they had covered about 120 miles, or about half-way back to their intended starting point. They entered Umivik Fjord as Nansen decided if they didn't depart soon the weather would make the journey extremely difficult if not impossible.

So after a few days of rest and preparation, they set out on a course for Christianhaab (now Qasigiannguit), on the west coast of Greenland, about 370 miles away. To say the least, the journey was not easy, safe or particularly pleasant. The weather was described as generally bad with violent storms and almost continuous rain. They had to ascend treacherous terrain with many deep hidden crevasses in the ice in the middle of lousy weather and freezing cold. Doesn't sound like a picnic in the park.

After almost two weeks of tough sledding, Nansen realized that their northwesterly path was too far to allow them to make it to Christianhaab by mid-September when the last ship was due to leave. Always the objective realist, and perhaps contemplating mutiny from his men, he decided to alter their plan and head almost due west to Gothaab (now Nuuk), which was about 93 miles closer and, importantly, at least remotely attainable.

By early September they had climbed to the summit of the Greenland icecap at an elevation of almost 9,000 feet above sea level, where temperatures dropped to 50 degrees below zero at night. But they had those great homemade sleeping bags to keep them warm.

The worst was over, although it had taken them about four weeks to get this far. Pulling their sledges was considerably easier now that they were heading downhill, and by late September they reached the head of the fjord along the southwest coast of Greenland. Nansen managed to construct a boat of sorts out of some local willows, parts of their sledge and their tent. They then rowed down the fjord to Gothaab, arriving four days later on October 3, 1888.

The first crossing of Greenland had taken 49 days and they had collected careful meteorological, geographical and other observations throughout their expedition.

They were greeted by the town's Danish mayor who informed Nansen that his Ph.D. committee had approved his thesis and awarded him his degree. This was an unexpected surprise, but balanced somewhat by the news that the last boat had just left and the next vessel wouldn't arrive until the next spring, seven months later.

Article 7

Nansen Heads for the Arctic

Fridtjof Nansen's Fram expedition bound for the Arctic, 1893.
(from Nansen's book, Farthest North (1897), via Wikipedia Commons)

Fridtjof Nansen and his small group of not so merry men were the first to cross the Greenland ice cap. When they arrived on the west side at Godthaab in October 1888, they discovered that the last ship for the year had just left and the next wouldn't arrive until spring. He and his party just made the most of it and spent the next seven months hunting, fishing, and studying the life of the local inhabitants, which they quite enjoyed.

They received a hero's welcome back in Norway in May 1889, and the enthusiasm for their achievement led directly to the formation of the Norwegian Geographical Society. The Royal Geographical Society of England awarded Nansen the Society's prestigious Founder's Medal, stating that Nansen had reached "the foremost place amongst northern travelers".

On a more personal side, while our young adventurer had made it no secret that he himself was strongly against the institution of marriage, less than three months after returning to Norway he was engaged, and then, less than a month later, Nansen married Eva Sars, the daughter of a zoology professor.

His successful Greenland crossing, combined with his spirit for exploration and long interest in the Arctic led him to think again about the possibility of reaching the North Pole through the natural movement of the ice. In February 1890, Nansen addressed a meeting of the new Norwegian Geographical Society where he proposed a plan to drift in a vessel across the Arctic.

In contrast to the earlier Arctic voyages of others that ended in disaster, and based on the theories of a distinguished Norwegian meteorologist, Nansen believed that success required that any ship drift with the direction of ice flow. This meant sailing from the Siberian side, where the Jeannette had been crushed in the ice in 1881, and then drifting towards Greenland. Nansen stated that the obvious thing to do was "to make our way into the current on that side of the Pole where it flows northward, and by its help to penetrate into those regions which all who have hitherto worked against (the current) have sought in vain to reach".

He planned to sail to where the Jeannette sank, and at the time of minimum ice extent, "we shall plough our way in amongst the ice as far as we can". He believed that the ship and crew would then drift with the moving ice towards the North Pole and eventually reach the open sea between Greenland and Spitsbergen.

As with his proposed Greenland trip, there were a number of polar explorers who were completely dismissive of Nansen's plan, including a retired American adventurer who called the idea "an illogical scheme of self-destruction". Another American explorer called it "one of the most ill-advised schemes ever embarked on" and predicted that it would end in disaster. But Nansen's earlier success and a passionate request to the Norwegian parliament led to a grant, followed by funding from private individuals and also support from a successful national appeal (which may very well have been the first example of crowdfunding).

He was on his way to his next great adventure and just needed the right ship and a trusting and dedicated crew who didn't mind spending five years stuck in the Arctic ice. Nansen's plan required a small, strong, and maneuverable ship, powered by both sail and an engine, capable of carrying fuel and provisions for twelve men for five years. Well, and then there was also the matter of several dog teams to pull the sledges in case they were needed, so a lot of dog food.

Nansen knew he was risking his and his crew's lives so went with quality. He choose Colin Archer, Norway's most respected naval architect and shipbuilder, to design and build a ship that could resist all of the forces the Arctic ice would throw at it. Archer built a vessel of extraordinary strength by using the strongest oak timbers he could find, combined with an intricate system of crossbeams and braces along the entire length of the vessel. The hull was sheathed in South American greenheart, the hardest timber available. Three layers of wood forming the hull provided a combined thickness of between 24 and 48 inches.

Sleek would not have been a good description of the ship. It was designed with a rounded hull so that as the ice froze around it, the ship would be pushed upwards onto the surface rather than being crushed in an ice vice of considerable pressure. In Nansen's words, the vessel would "slip like an eel out of the embraces of the ice". The vessel was also quite broad, 36 feet, relative to its length (128 feet), giving it a somewhat stubby appearance, but then beauty wasn't a criterion.

On the 6th of October 1892, the ship was launched, christened by Nansen's wife Eva, and given the name Fram, meaning Forward, because "forward" was the only direction Nansen was willing to consider.

One might think that there wouldn't have been much appeal of a five-year voyage in a small ship, stuck in the Arctic ice with 12 other men and several dozen dogs. But Nansen received thousands of applicants from all over the world. In the end, he selected only Norwegians and carefully handpicked each crewmember for their skills and strengths. Second in command was Otto Sverdrup, an experienced sailor who had accompanied him on the Greenland crossing.

Before setting off, Nansen made the decision to alter his original plan. Instead of following Jeannette's route through the Bering Sea to the New Siberian Islands, he decided to take a shorter route around the top of Norway and then proceed eastward along the northern coast of Siberia.

The Fram departed Christiania on June 24, 1893 with thousands cheering them on. The last port call was in Vardo, on the far north coast of Norway, and from there they headed east along the coast of Siberia. These seas were little known and poorly charted and they were slowed by ice and heavy fog, even in mid-summer. Today, with the rapid melting of the Arctic ice, their voyage would have been considerably easier.

As they approached the New Siberian Islands, where Jeannette had been caught in the ice and crushed, they began to encounter heavy pack ice. On about the 20th of September 1893, they had reached 78° 49′ N latitude before Nansen ordered the engines stopped and the rudder raised. From this point, the drift of the Fram in the ice began and would last for the next three years.

Article 8

Drifting Towards the North Pole

Fridtjof Nansen's ship Fram, frozen in Arctic ice, 1894.
(from Nansen's book, Farthest North (1897), via Wikipedia Commons)

On the 20th of September 1893, the Fram with Fridtjof Nansen and his crew of 12 and about two dozen sledge dogs were frozen into the ice north of the New Siberian Islands at latitude 78° 49'. If his ideas were correct, and with any luck, the ship with the crew and dogs would drift along with the Arctic ice, pass near the North Pole, and several years later, emerge near Greenland.

On the other hand, if Nansen was wrong or if any of a number of things didn't go as planned, they would all die. They didn't really have a lot of options once the ice froze around the ship's hull. But then Nansen had christened the ship Fram, which meant forward in Norwegian, and that was the only direction he planned on going.

The first several weeks were disheartening as the ship moved with the ice in a circular fashion, forwards and then backwards. After six weeks they were further from the pole than when they had begun. The lack of forward progress and boredom were hard on the men but Nansen did his best to keep them occupied with scientific observations and other tasks.

Finally, in January 1894, they began to move steadily northward and crossed the 80° latitude mark on 22 March. Nansen determined that at this rate it might take as long as five years to reach the pole. He and Hjalmar Johansen, an expert dog-driver, began to privately discuss the possibility of making a sledge journey to the pole if conditions didn't improve.

Unfortunately, Nansen had never driven a dog sledge before and his initial attempts were somewhere between humorous and embarrassing. But he persisted and before long his skills improved. He also made an important discovery, that men on cross country skis could keep up with the dogs and therefore didn't need to ride on the sledges. This allowed the sledges to carry more supplies.

A month later the Fram passed 81° north, which while encouraging, their progress still only amounted to about a mile per day. The sledge expedition to the pole was announced to the crew with the intention to depart when the ship passed 83°. Sledges were built that would allow for travel over the rough ice terrain, and kayaks were also constructed for the time where they might have to enter the sea.

Nansen and Johansen intended to reach the pole, then head south to Franz Josef Land and then on to Spitzbergen where they hoped to find a ship to back to Norway. There were lots of uncertainties but these guys were as well prepared as they could have been, and between them had a lot of Arctic experience. Unfortunately neither GPS nor cell phones had yet been invented, so they faced some formidable challenges just knowing where they were going to be.

Article 9

Nansen's Dash to the Pole

Nansen prepares to leave the Fram to journey by sledge to the North Pole, 1895. (from Nansen's book, Farthest North (1897), via Wikipedia Commons)

On March 14, 1895, after 18 months in the ice, Nansen and Johansen, who was fortunately the most experienced dog team driver on the crew, were ready to make their dash to the pole. They had built the sleds and kayaks, and had gone through endless machinations about what would be the best gear and supplies to take.

Two months earlier, the entire expedition was thrown into serious doubt as the Fram began to tremble violently when the ice squeezed the hull of the ship. Nansen had the entire crew leave the ship, fearing that it was finally going to succumb to the pressures of the ice. But the thick hull and internal bracing held, they all climbed back on board and preparations for the sled journey continued.

On February 17, Nansen had began a farewell letter to his wife, Eva, writing that should things not work out that "you will know that your image will be the last I see". He was also reading about Franz Josef Land, their intended destination after reaching the pole. It had been partially explored in 1873, and there were apparently large numbers of both bears and seals, which Nansen realized might come in handy to sustain them if they got that far.

Nansen and Johansen had two earlier departure attempts in late February but problems with the sleds required their return to the ship to overhaul their equipment and reduce their loads. They finally settled on three sleds, which seems a little challenging considering there were only two of them. But off they went to the North Pole, 410 cold, rough, ice covered miles away.

Their first week went smoothly and optimism was high as they traversed flat snowfields. Nansen calculated that they would need to average about 8 miles a day to make the pole in 50 days, which was about the amount of food they could carry for the two of them and the dog teams. After eight days their sextant observations indicated that they had been speeding along at 10 miles a day. They were pleased with their progress, particularly with temperatures averaging 40 degrees below zero.

Conditions then began to deteriorate as the surface of the ice got rougher, slowing their progress over the next week. They had also lost a device used to measure how much distance they had covered across the ice. Other measurements suggested that while they were moving north, the ice beneath them was actually drifting south, slowing their progress considerably.

Nansen's journal entry at this point suggested some serious doubt about their outcome: "My fingers are frozen stiff... it is becoming worse and worse... God knows what will happen to us".

On April 3, 1895, after nearly 20 days of tough sledding, Nansen began to wonder if they would reach the pole and still have the food they needed to get to Franz Josef Land. In his journal he wrote "I have become more and more convinced we ought to turn before time". Four days later, on April 7 at latitude 86 degrees, 13.6 minutes north, the farthest north that humans had ever reached, with a chaotic mass of large ice blocks extending to the horizon, he decided that in order to survive, that they would have to turn back and head towards Franz Josef Land.

Article 10

Trying to Find Franz Josef Land

While they were no longer heading forward, after turning away from the North Pole and heading south, Nansen and Johansen initially made good progress as they encountered areas of relatively smooth ice. Their enthusiasm was high until April 13, 1895, 7 days into their trek to Franz Josef Land, when both of their watches stopped.

This set the stage for potential disaster. Without knowing the time, they couldn't calculate the longitude correctly and, therefore, had no way to navigate accurately. Nansen made a guess of the time and they reset their watches, but they weren't at all certain of their true position. If they missed Franz Josef Land, it was probably all over for the pair.

But as they progressed with the dogs across the ice, feeling considerable uncertainty about their future and ultimate destination, things started to brighten up a bit. Near the end of April, now six weeks after leaving the ship, they saw tracks of an Artic fox, the first trace of another living creature beyond their dogs since they left the Fram. They soon noticed bear tracks, and then by the end of May, seals, gulls and whales in the occasional open water. At least they weren't going to starve to death.

At that point, by Nansen's calculations, they were only about 48 miles from the northernmost point of the Frans Josef Land islands. But the ice was starting to break up, making progress across the ice more difficult. Since late April they had been killing dogs at regular intervals to feed the others, and by the beginning of June only seven of the original 28 remained. On June 21, they left behind all of the supplies and equipment they felt they could do without in order travel as light as possible, living off the now plentiful supply of seals and birds.

The next day they decided to camp on a large ice flow, repair their gear, waterproof their kayaks and collect some strength for what they hoped would be the final stage of their expedition. A month later, on July 23, 1895, the day after leaving their camp, they saw land for the first time in almost two years. In his journal Nansen wrote: "At last the marvel has come to pass - land, land, and after we had almost given up our belief in it".

They survived a polar bear attack, and struggled across the last few miles of ice to get to an area of open area of water where they finally heard surf breaking. On August 6 they converted the kayaks into a catamaran, put up a sail and headed for the land they were so happy to finally see.

They soon observed that the land they had encountered was part of a larger group of islands. Moving southward in their kayaks, with an earlier map in hand, Nansen tentatively identified a headland as Cape Felder, which led them to believe they had in fact found Franz Josef Land. While they had hoped to find a hut and supplies further south left by an earlier expedition, loose ice and unfavorable winds made that impossible.

So although it was only late August, the Arctic winter was beginning to close in on them and they decided to camp for the winter. In a sheltered cove, with rock and moss for building materials, they constructed a hut, which was to be their home for the next eight months. Their situation was uncomfortable, but not life threatening. There was plenty of wildlife around but their main enemy over the next eight months was to be boredom. They read and reread Nansen's sailing almanac and navigation tables by the light of an oil lamp.

Article 11

Nansen Trying to Get Home

Fram expedition members upon their return to Christiania, 1896.
(from Nansen's book, Farthest North (1897), via Wikipedia Commons)

We left Nansen and Johansen in a crude hut somewhere near Franz Josef Land where they were to spend eight long months together before emerging like bears from hibernation on May 1896, ready to resume their journey.

Meanwhile, the Fram with its crew was still stuck fast in the ice far to the north. Sverdrup had been left in charge and had his hands full after two long years with a bored crew surrounded by nothing but ice. He kept the men busy, cleaning up the ship, chipping off ice, organizing provisions in the event they had to abandon ship, taking short ski expeditions and collecting scientific observations. The ship had become a moving oceanographic, meteorological and biological laboratory.

Soundings indicated deep water and no Arctic landmass. After months of moving towards the pole, they began to drift southward after having reached within just 19 miles of Nansen's most northerly point. After another long winter, leads in the ice began to open, and finally, on August 13, 1896, almost three years after being first frozen in, the Fram left its icy prison. The ship had emerged from the ice just northwest of Spitsbergen, very close to Nansen's original prediction.

The Fram sighted another ship that same day, a seal hunter from Spitsbergen, but there had been no sighting of Nansen and Johansen, so Sverdrup and his crew began the trip south and east back to Norway.

Nansen and Johansen followed the coastline southward although they were a bit distraught that nothing seemed to match the rudimentary map they had. As weather improved they decided to launch the kayaks again, although at one point, Nansen had to dive into the icy waters as their boats had drifted away due to a careless mooring. He managed to reach the boats and with a final last ditch effort, managed to haul himself aboard.

On June 13, 1896 walruses attacked their kayaks causing a delay for repairs. Four days later, as they were getting ready to launch their crafts again, Nansen thought he heard a dog bark. As he went to investigate he heard voices and then later saw a person.

It was Frederick Jackson, who had organized an expedition to Frans Josef Land after being rejected by Nansen as a crewmember on the Fram. As he approached, Jackson saw "a tall man, wearing a soft felt hat, voluminous clothes and long shaggy hair and beard, all reeking with black grease". After a moment's hesitation, Jackson recognized the man: "You are Nansen, aren't you?" and received the reply "Yes, I am Nansen".

The pair was rescued after having survived for 15 months alone in the middle of the Arctic in less than pleasant conditions. On top of that, between them they had gained 34 pounds while in hibernation. They patiently waited six more weeks for the arrival of a ship, the Windward, which would take them back to Norway.

On August 13, 1896, Nansen and Johansen arrived at the port of Vardo. A flurry of telegrams was sent announcing to the world that they had returned safely. Nansen was concerned, however, that he had heard nothing about the fate of the Fram. Within a week, however, he received the news that Sverdrup had brought the ship to the small port of Skjervoy, just south of where he and Johansen were. The very next day, Nansen and Johansen sailed to meet their comrades for a very emotional reunion.

Although it did not achieve the objective of reaching the North Pole, the expedition made major scientific and geographical findings. It was now established that the North Pole was not located on land, nor on a permanent ice sheet, but on shifting, unpredictable pack ice. The Arctic Ocean was, in fact, a deep basin. From its careful and sustained scientific observations the voyage of the Fram provided the first detailed oceanographic information from the Arctic.

The Fram voyage was Nansen's final expedition and the ship is still on display in the museum in Oslo. Nansen was just 35 years old and was soon appointed as a research professor at the University of Christiana in 1897. In his later career he served the newly independent kingdom of Norway in many different capacities, and was awarded the Nobel Peace Prize in 1922, in recognition of his work on behalf of refugees globally. He lived an adventurous and very full life.

PART X

CLIMATE CHANGE

ARTICLE 1

Climate Oscillations and Disappearing Sardines

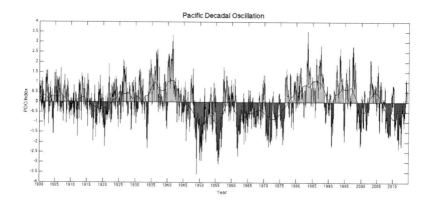

Pacific Decadal Oscillation
(courtesy: Joint Institute for the Study of Atmosphere and Ocean, University of Washington)

A half century ago, the city of Monterey changed the name of Ocean View Avenue to "Cannery Row", in honor of John Steinbeck's 1945 classic story. The sardine fishery made Monterey one of the largest fishing ports in the world through much of the first half of the 1900's, but by 1945, the fishery had collapsed and Cannery Row as it existed then, went with it. The Monterey Bay Aquarium was opened in 1984 on the old Hovden Cannery site, using parts of the old cannery as exhibits that give us a glimpse of the past.

Following the collapse of the sardine fishery, which at its peak involved 200 active fishing vessels that brought in about 650 million fish each year, there were several theories or explanations as to what caused this failure in the fishery. Foremost was perhaps the use of purse-seiners, the "Wolves of the Sea", with nets a quarter of a mile long that reached 200 feet down and simply over-fished the population down to a point where the fishery wasn't sustainable. We have, unfortunately, now done this with many other fisheries as well, the cod, and many of our rockfish, for example. Another argument was that the increasing use of pesticides such as DDT, on the crops of the Salinas Valley, had impacted Monterey Bay by runoff and biological magnification up the food chain, which had seriously impacted the sardine population.

In subsequent years, the careful analysis of sediment cores collected from the sea floor off the southern California coast, as well as off Peru, revealed that there had been intervals of time, hundreds of years ago, well before commercial fishing was initiated, when sardine scales were common in the seafloor mud; during other time intervals the scales were virtually absent. It soon became apparent that sardine populations come and go over periods of several decades due to some unknown cyclical ocean conditions.

An early ocean observation system was set up in 1949, known as CalCOFI (the California Cooperative Oceanic Fisheries Investigations), to study the ecological aspects of the collapse of the sardine populations by regularly measuring such ocean variables as temperature and salinity in the offshore California Current.. These measurements and many others are still being collected today (such as the CenCOOS ocean observation program, mentioned in another column) in order to provide a longer-term perspective on natural cycles and to help us answer these sorts of questions.

What we have now discovered from these long-term observations is that the ocean climate in the Pacific varies over time periods of several decades in what is now known as the Pacific Decadal Oscillation (PDO). Changes in ocean temperatures over large areas affect atmospheric pressures and subsequently wind patterns. The storm climate, ocean surface temperatures, as well as the intensity of upwelling and the availability of nutrients, all change in cycles that may last 20 or 30 years. It is these changing ocean conditions that influence which species thrive and which decline in abundance. So sardines come and go in cycles that may last 25 or 30 years, and in Monterey Bay they seem to alternate with anchovies. From about 1920 to 1945, in what is now known as a warm phase of the Pacific Decadal Oscillation, the sardines dominated, the fishery expanded and the number of canneries increased. Cannery Row was in its heyday.

In 1945 the climate shifted to the cool phase of the PDO; the sardines disappeared, replaced by anchovies. This cool phase continued until about 1978 when the sardines returned again, although the market for sardines was much different than it had been 30 years earlier. The period from 1945 to 1978 was also generally characterized by fewer large storms, cooler ocean water, lower rainfall, and less coastal storm damage. This all changed along the central coast in 1978 when we entered a period of more frequent and damaging El Niño years, which stayed around until the early 2000s.

ARTICLE 2

What's Next For Our Shoreline? El Niño and La Niña

Impact of a harsh El Niño on Seacliff State Beach, Santa Cruz County, 1983.
(© 1983 Gary Griggs)

It's always dangerous to say you are going to write about something in your next column, because if you are to maintain any credibility, you had better finish the story. Two weeks ago I wrote about our relatively recent awareness of climate cycles and shifts and how we believe that these were reflected in the history of the sardine fishery in Monterey Bay. These climate changes that are now known to affect the entire Pacific Ocean also have had profound impacts on the California coastline.

The sardines had more or less disappeared by about 1945 as we entered a period generally characterized by cooler temperatures, less rainfall, fewer severe coastal storms, and reduced frequency and intensity of El Niño events. This interval, also known as a cool phase of the Pacific Decadal Oscillation (PDO), lasted until about 1978. It also just happened to coincide with the years following World War II when people migrated to California in huge numbers and the coastline of southern and central California was rapidly developed.

The beaches were wide, the coastline was appealingly warm and dry, and California seemed much more desirable than the freezing winters of the mid-west and the summer humidity of the east coast. Houses were built as close as possible to the shoreline. We built on cliffs, on dunes, and on the beach itself, year after year. The limited amount of coastal property became increasingly more valuable. In Malibu, one of the desirable ocean front addresses was Sea Level Drive. On Del Monte Beach in Monterey we have Wave Crest and Surf Way, both suggestive to me of occasional exposure to high tides and large waves.

This ideal oceanfront world was rudely awakened with the arrival of the winter of 1978 when the first major El Niño in decades pummeled the California coast, including Monterey Bay. Low-lying areas like the Esplanade in Capitola, Pot Belly Beach, Seacliff State Beach, and Beach Drive in Rio Del Mar were hammered with a combination of El Niño driven elevated sea levels, high tides and large storm waves that arrived from the west and southwest, hitting the normally protected shoreline of northern Monterey Bay. The beaches were eroded and the waves then started in on landscaping, decks, foundations, and then picture windows and sliding glass doors that overlooked the beach. The real estate ads describing homes that were "on the sand", took on new meanings.

People in California tend to have a collective amnesia for events like this, so within a few months when the winter storms had passed, the beaches returned, insurance companies were notified and repairs were made. In some cases, coastal homes were sold to unsuspecting new arrivals. 1978 was followed by the memorable El

Niño winter of 1983, perhaps the most damaging to the California coast in half a century. Much of the newly reconstructed timber seawall at Seacliff State Beach was destroyed (for the 8th time in 60 years), waves washed into the Venetian Court Condominiums and the restaurants on the Esplanade in Capitola, and homes along Beach Drive in Rio Del Mar suffered major damage as seawalls were destroyed, sand levels dropped and some houses collapsed onto the beach.

Impact of 1983 El Niño on Rio Del Mar Beach, Santa Cruz, County.
(© 1983 Gary Griggs)

At Aptos Seascape, the existing protective rock revetment was overtopped and the storm waves went through the sliding glass doors into living rooms and kitchens. Damage was extensive. Ocean front houses at Pajaro Dunes were threatened as 40 feet of sand dune was eroded almost overnight leading to the emergency placement of 1000's of tons of rock in an effort to save the homes. Following the winter of 1983, the Coastal Commission and County Planning Department were inundated with permit requests for new seawalls and more riprap. The storms continued intermittently through the 1980's and 1990's, with the El Niño of 1997-98 being another major damaging event along our coast. By this time, many of the threatened properties had been sold or armored with seawalls or riprap.

What can we expect next? It appears as though we have transitioned from the warm PDO phase with its severe coastal storms that extended from 1978 to perhaps 2000, and have entered a cooler and calmer Pacific Decadal Oscillation phase. At the same time, however, indications are that the rate of sea-level rise has increased over the past 20 years and as global climate change continues, it isn't clear yet how this may affect the frequency and magnitude of future El Niño events.

ARTICLE 3

Calera-Green Cement for a Blue Planet

Making cement from seawater? Hearing this for the first time took me by surprise, but after talking to Brent Constantz, who started a company that is doing just that, it made more sense. This endeavor could help resolve a serious problem, that of the impact of the growing carbon dioxide content of the atmosphere on global climate. Brent, who received his Ph.D. in Earth Sciences from UCSC, started this company a few years ago as he became more concerned with the impact of increasing CO_2 emissions.

Brent's graduate research focused on corals. He was intrigued with how they made such strong skeletons from seawater. Brent's career took him to Stanford where he became a Consulting Professor and began translating his knowledge of how marine organisms formed their skeletons into developing cements for mending broken bones. When he was 27 Brent created a cement that revolutionized the repair of broken bones in hospitals around the world, based on his work on corals.

Concrete is the world's most widely used building material, with about 12 billion tons of it used each year. The primary ingredient of concrete is cement, which is produced from limestone. Limestone, or $CaCO_3$ is just the accumulation of the remains of millions of marine organisms that made their shells out of a few basic elements in seawater like corals do. The limestone (now metamorphosed to marble) under the Westlake neighborhood, under parts of UCSC and also Bonny Doon, has provided the raw material for the county's cement industry for over a century.

At the former cement plant at Davenport, limestone was heated up to over 1800° F., which drives off CO_2 and leaves lime or calcium oxide, the principal ingredient in cement. This process, however, releases 1 ton of carbon dioxide into the atmosphere for every ton of cement produced. Worldwide, about 2.5 billion tons of cement is manufactured each year, emitting 2.5 billion tons of CO_2, about 5% of the world's total CO_2 emissions. Brent Constantz thought there might be a better way to make cement, and started a company called Calera (lime kiln in Spanish).

After months of laboratory scale work in Los Gatos, hauling Monterey Bay seawater from Long Marine Laboratory over the hill, Calera's team of chemists and engineers developed a technique for producing small batches of cement from the seawater. After initial successes they scaled up the operation and set up shop in the old Kaiser Refractories plant at Moss Landing, which produced material for bombs during World War II, but extracted magnesium from seawater for various industrial uses in post-war years.

Right across the street from the plant is the largest power plant on the west coast, which burns natural gas and then releases carbon dioxide into the atmosphere from its enormous stacks. Brent and his team are now well along on the technology that can remove about 90% of the carbon dioxide from the flue gases by bubbling the exhaust from the power plant through the seawater while they are making the cement. For each ton of cement they make, they can incorporate a half a ton of carbon dioxide into the cement. By building cement plants on the coast next to fossil fuel fired power plants, Calera can solve two huge challenges, making green cement by not producing additional CO_2, and also removing the CO_2 from fossil fuel burning by sequestering it into concrete.

Article 4

Climate Change and Rising Sea Level

The preserved records of the climate change that the Earth has experienced throughout its history are diverse, widespread and well documented. These include the extent of past glaciers and the stuff they left behind, the isotopic signatures from long ice cores from Antarctica and Greenland (indicating alternating warm and cooler periods that go back as far as 800,000 years), and the fossil record from deep-sea sediment cores that extends back many millions of years.

Climate change has taken place ever since we have had an Earth and a sun, and in fact, the climate is always changing. These fluctuations were first discovered over 150 years ago and correctly attributed in large part to the irregularities in the Earth's orbit around the sun and how much heat we receive.

So why is the climate change taking place today such a big deal? Well, for one thing, while there have been much warmer periods in the geologic past, as well as much cooler periods, there weren't any or many people around to deal with those extremes. Mastodons and cave bears came and went. Entire populations of plants and animals flourished, migrated or disappeared. When it got warm and the ice melted, sea level rose and the coastline just moved 10, 20, or 50 miles inland. But we didn't have cities on the shoreline like Santa Cruz, San Francisco, New Orleans, or New York City to deal with 125,000 years ago. Climate change happened; it happened repeatedly, and whatever was around adapted, migrated or simply went extinct.

Another important factor is that the 7.1 billion people now on Earth have had a very measurable effect on the Earth's atmosphere.

Greenhouse gases have been accumulating and increasing for 150 years and denial is not a river in Egypt.

The most recent state, national and international reports on climate change and sea level conclude that the most likely scenario would lead to a 1 meter or roughly 3 feet of sea-level rise by 2100. It might be more and it might be less. Several years ago, the San Francisco Bay Conservation and Development Commission (BCDC) was concerned enough about the potential impacts of sea-level rise on the low-lying areas around San Francisco Bay that they commissioned a survey to map out what might be threatened in the future. A one-meter rise in sea level would completely inundate the San Francisco and Oakland International airports, as well as sections of the Bayshore Freeway. This would present some significant challenges.

John Holdren, a highly respected physicist from Harvard, and the Science Advisor to President Obama, has stated in regard to global climate change that "We basically have three choices: mitigation, adaptation, and suffering. We're going to do some of each. The question is what the mix is going to be. The more mitigation we do, the less adaptation will be required and the less suffering there will be."

Mitigation suggests that there is something we can actually affect or alter. Can we mitigate climate change by greatly reducing our greenhouse gas emissions? Can we adapt or respond to climate change, or are we, and our children, just doomed to suffer? Until somewhat recently, climate change seemed to be happening fairly slowly, but many indicators are showing more rapid change than most scientists had predicted. We should be hoping for the best but planning for the worst.

ARTICLE 5

Messing with the Atmosphere

One thing that scientists agree on today is that global climate has constantly changed throughout Earth history, and that key drivers are the oscillations in the Earth's rotation on its axis and in its orbit around the sun. These cycles are well understood and predictable. They take place over tens of thousands of years and are way beyond human control. They just are.

There is also no scientific disagreement that sea level has risen and fallen repeatedly throughout geologic time in response to climate change. Sea level just happens to be at a high point right now simply because we are in a relatively warm climate phase and much of the Earth's ice has melted. But sea level has been considerably lower in the past and it has also been a lot higher.

The question that often generates the most discussion is how much human activity is affecting climate change, which directly affects sea level. When it gets warmer, glaciers and ice sheets melt and sea level rises. There's a very direct connection. There is no other place all of that water can go.

A half-century ago several scientists at the Scripps Institution of Oceanography had the suspicion that our burning of increasing amounts of fossil fuels, primarily coal, oil and gas, could be affecting the carbon dioxide content of the atmosphere, simply because burning any fossil fuel releases carbon dioxide. One of those scientists, Charles Keeling, went to the summit of Mauna Loa Observatory in Hawaii where he felt the air would be uncontaminated by industry and urbanization, and began to measure the amount of carbon dioxide. While most of our air consists of nitrogen and oxygen, he found when he arrived in Hawaii in 1958 that the carbon dioxide concentration was 315 parts per million (ppm). Year after year Keeling continued these annual measurements, and each year the concentration of carbon dioxide increased.

Charles Keeling died in 2005, but during his 48 years of continuous the carbon dioxide concentration had increased 20% to 380 ppm. His son, Ralph Keeling, a climate scientist at Scripps took over his father's measurements. Carbon dioxide in 2013 reached 400 ppm, higher than any time in the 55 years of direct measurements in Hawaii, and probably the highest it has been anytime in the past 3 million years. Many scientists credit Keeling's careful measurements with first bringing the world's attention to the effects that human activity were having on the Earth's atmosphere.

But is 50 years really long enough to get a representative record? In order to get a longer-term picture of the significance of these changes, scientists have now taken measurements of the carbon dioxide concentration in ancient air bubbles trapped in polar ice cores. These analyses show that average atmospheric CO_2 concentration varied between 275 and 280 ppm throughout the past 9000 years, but started rising sharply at the beginning of the nineteenth century.

The CO_2 content of the atmosphere has increased about 43% over the pre-industrial levels and is now well beyond the range ever experienced in the entire time that humans have existed on the Earth. The increase is considered to be largely due to the combustion of fossil fuels. Since carbon dioxide is a greenhouse gas, meaning it tends to trap heat, this has significant implications for global warming. There is a natural greenhouse effect, which makes the Earth a pleasant place to live, but we have now added an additional greenhouse effect, and therein lies a problem.

Article 6

Greenhouse Gases and Climate Change

Fortunately for all of us, the Earth's atmosphere contains natural greenhouse gases that have made the Earth a habitable planet by keeping it warmer than it would be otherwise. Our atmosphere behaves much like the windows in your car. Short wave solar radiation passes through the glass. The interior heats up and reradiates long wave back radiation. But the glass windows don't allow it to escape, so the inside of your car can get pretty hot.

The Earth works much the same way, only instead of glass we are surrounded by an atmosphere that naturally contains small amounts of greenhouse gases, primarily carbon dioxide, methane, nitrous oxide and fluorocarbons. These greenhouse gases trap most of the long wave back radiation, increasing the temperature of our atmosphere and making life bearable. The average surface temperature of the Earth is about 50° F (or 15° C), but without the natural greenhouse gases it would be a lot colder, in fact, just about freezing.

But, we've been seriously messing with the atmosphere for well over a century. In the past 150 years, burning fossil fuels, deforestation, and the production of cement have increased the amount of CO_2 by about 43%. Methane, although only about 1/6 as abundant as carbon dioxide, can trap about 20 times more heat. It is generated by livestock, coal mining, rice cultivation, oil drilling and landfills, and has more than doubled in concentration since 1850.

Instead of rolling down your car windows an inch to let some of that heat escape, adding more greenhouse gases to the atmosphere is like rolling up the window a little tighter. The Earth is slowly getting warmer, and there are signs of this nearly everywhere you look across the planet. Perhaps the most visible are centuries old ice shelves breaking up in Antarctica and the reduction in the ice cover of the Arctic Ocean. While Australians have routinely weathered dry spells, a recent seven-year drought was the most devastating in the country's 117-year history.

Some have questioned whether the human activities just described are capable of producing the changes we are now observing. Can this warming be due to natural patterns? While the Milankovitch cycles in the Earth's rotation and orbit can explain climate variations over very long time scales (tens of thousands of years), these cycles do not explain variability in climate on the scale of decades we have been experiencing. Variability over shorter time periods is driven by other factors, such as CO_2 and other greenhouse gas concentrations. We cannot produce the warming patterns we have been measuring over the past 150 years by relying solely on variations in solar radiation.

Measurements of carbon dioxide concentrations in gas bubbles preserved in ice cores recovered from Antarctica indicate that the amount of CO_2 in the atmosphere today is higher than anytime in at least the last 800,000 years. We have had a very significant impact on the composition of the atmosphere.

I suppose there are two possible scenarios at this point: either the climate changes we are beginning to experience and the effects these will have on our lives are all natural and there is absolutely nothing we can do but suffer and learn to adapt; or the changes we are experiencing are due primarily to the accumulation of greenhouse gases, and by substantially decreasing those we can over time begin reduce our suffering. Several years ago the Environmental Protection Agency declared CO_2 and five other heat-trapping gases pollutants that endanger public health and welfare. In June 2014 the Supreme Court upheld President Obama's plan to cut carbon dioxide emissions from power plants.

Article 7

Collapsing Cliffs and Shifting Climate

Erosion continues to undermine Pacifica apartments, 2013.
(© 2013 Gary Griggs)

February 2010 was a bad month for the residents of an apartment complex on a high sandy bluff in Pacifica. They had to evacuate their homes as winter storm waves continued to erode the bluff fronting their two-story apartment building. Chunks of bluff continued to fail until their narrow concrete patios were left dangling over the edge, 80 feet above the ocean. In addition to dumping a half a million dollars worth of rocks at the base of the bluff in an attempt to halt wave attack, crews from a construction company began a $6 million bluff stabilization project. The plan was to build a soil nail wall, similar to what was been constructed along the low bluffs at Pleasure Point. I had stopped by the site on my way back from a meeting in San Francisco. Workers in a metal basket suspended from a huge crane were using a portable drilling rig to insert tiebacks, or huge nails, about 50 feet into the bluff. These soil nails will be grouted or cemented into place and then steel plates and a wire mesh will be connected to the soil nails, which will then be covered with shotcrete or spray on concrete.

The head of the construction company says he is dead certain that their proposed engineering fix will stop Mother Nature in her tracks - at least for 50 to 70 years. As an update, less than three years later this "engineering fix" was all but gone. Looking at the loose sandy materials making up these bluffs, I asked myself why a series of large apartments and other buildings were built here to begin with? A few hundred feet to the south along the Esplanade an entire row of older ocean front homes were undermined and were either demolished and collapsed onto the beach just a decade ago during the 1997-98 El Niño winter. A half a mile in the other direction is the San Andreas Fault, which hasn't ruptured since 1906, but strain has been accumulating for well over a century. This is not the place to invest in ocean view property.

These high sandy bluffs have been eroding for decades. Long-term measurements of cliff retreat from aerial photographs indicate average erosion of a foot to a foot and a half a year. Some sections of this bluff to the south were cut back 30 to 45 feet during the 1982-83 El Niño winter alone. Mobile homes actually became mobile, and were moved inland as their concrete pads were undermined and existing seawalls were destroyed.

The Daly City and Pacifica area were urbanized between the 1950's and the early 1970's, like much of California's coast, with construction encroaching closer and closer to the edges of the bluff. We now know that the storm and wave climate of the Pacific Ocean oscillates or fluctuates over cycles typically lasting several decades. These Pacific Decadal Oscillations exert a strong influence on the coastline by virtue of changing storm conditions and wave energy.

The period from about 1945 to 1977 was a cool or La Niña dominated period characterized by less intense winter storms, lower rainfall and less severe wave attack of the coastline. This was also the period following World War II when California's population exploded and development took place on the cliffs, bluffs and beaches of California. This all changed in 1978, however, when the coastal climate shifted to an El Niño dominated period, bringing elevated sea levels, intense storms and coastal wave attack.

Article 8

Santa Cruz-Going to Extremes

Its often been said that climate is what you predict and weather is what you get. Actually, climate is large scale and long-term, while weather is the day-to-day conditions and variations we experience in our own backyard.

We have a Mediterranean climate along the central coast, with the adjacent ocean serving as a buffer or climate moderator so that we're not usually nearly as warm or as cold as our inland neighbors. The winter of 2012 gave us some nice summer weather, however, and also broke some long-term temperature records. We don't usually get 80 degree days in March, but on March 4, the temperature in Santa Cruz peaked at 83 degrees, shattering the previous March 4 record of 79 degrees set in 1937, 75 years earlier.

The balmy weather extended north and south along the coast. In Southern California, the National Weather Service reported a record high of 95 degrees in Fullerton, 91 degrees in Long Beach and 87 degrees at the Los Angeles International Airport. Beaches from Santa Cruz to San Diego were crowded with people looking for a winter tan.

While it was hot that first weekend in March 2012, other winter days in the past have been even hotter. On February 23, 1896, it was 89 degrees in Santa Cruz. March has also experienced some very warm days. On March 17, 1914 it was 89 degrees and a day later the temperature reached 90 degrees-in March!

The hottest day ever in Santa Cruz took place on August 1, 1900 when the thermometer hit 108 degrees. But the warmest continuous weather occurred in late-June 1976, in the middle of a two-year drought. On June 23 the temperature hit 100 degrees, and for the next 5 days, maximums were 106, 105, 103, 105, and then it cooled to 103 on June 28. This is not typical or normal but it may well become more so in the future.

How about the other extreme? How cold has Santa Cruz ever gotten? Only twice since record keeping began in 1893 has the thermometer dropped below 20 degrees. On January 3, 1907, a low of 15 degrees was reached, and on December 23, 1990, the temperature dropped to 19.

Warmer climates are usually associated with lower rainfall, and California is no exception. While access to a reliable source of water was a critical factor historically in the development of civilizations and growths of cities, this has not been the case in recent time. Much of the west was replumbed by the Bureau of Reclamation and the Army Corps of Engineers to open up arid land for agriculture and to allow cities to expand where local water supplies were inadequate. Through a huge system of dams, reservoirs, pumps, pipes and canals, we built a system to move water hundreds of miles to where politicians and engineers at the time decided that it was needed.

Los Angeles ran out of local water in the 1880s and its subsequent search for water, the armed battles over the ownership of that water, and the construction of a delivery system to transport water to Los Angeles, has been well chronicled in both books and movies. The three million residents of San Francisco, San Mateo and Alameda counties obtain their water from the Hetch Hetchy reservoir in the Sierra Nevada, which was created by damming the Tuolumne River in the 1920s, amidst a large environmental outcry led by John Muir.

While large regions of California are completely dependent on imported water, Santa Cruz is somewhat unique in relying completely on our own water supply. We are hydrologically self-sufficient, with the only exception to this being the water we import in those cute little bottles at exorbitant cost and energy from Fiji and France.

Our local water availability, as a result, is dependent upon an adequate amount of yearly rainfall over the watersheds of the county. Rainfall is officially calculated by the USGS beginning October 1st in one year, and extending through September 30th in the next year. This way an entire winter of precipitation is counted within a single year. The city of Santa Cruz averages a little over 30 inches annually, although over the past 120 years this has varied from a low of just over 10 inches in 1898, to a high of 61.6 inches in 1941. While low rainfall years weren't a particular problem in 1898 when Santa Cruz had a population of about 5,600 people, it's a different issue today with a population ten times larger, and for a water system that now serves over 90,000 people.

California, as well as Santa Cruz, has experienced droughts in the past, and 2013-2014 has been the 4th driest year on record with 13.5 inches of total rain, and it follows a year with just 17 inches. The worst two-year dry spell over the period of historic record is probably still clear in the minds of many longer-term residents. During the 1976-77 drought, a total of 29.1 inches of rain fell in two years, just half of the average annual rainfall for two years in a row. 2013 and 2014 rank as the 2nd driest two-year period with a total of 30.6 inches. For the past three years (2012-2014) the city has averaged only 17.1 inches each year.

The longest sustained dry period occurred fairly recently, between 1987 and 1991; rainfall was well below normal for 5 years in a row, averaging just 19.5 inches annually. Between 1928 and 1934, the city had 7 below normal years in a row, averaging just 20.2 inches/year. Those were different times, however, and not only were there far fewer people but each person then used a lot less water.

Article 9

Changing Climates and Shifting Species

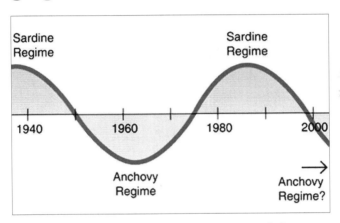

Alternating cycles of sardine and anchovy populations.
(© 2003 Monterey Bay Aquarium Research Institute)

This week marks two years of Ocean Backyard columns. When I began in April of 2008, I thought there might be enough ideas to last 6 months or so, but the ocean and our coastline continue to provide more ideas to write about. For those of you who email me from time to time, thank you for reading the columns, for your interest, and for posing questions and suggesting new topics. I think there are still a lot more issues and questions to explore and new ones continue to arise.

In an early column I talked about the rise and fall of Cannery Row. When the sardine fishery collapsed in the mid-1940's, there was no agreed upon answer as to why. With the beginning of regular and sustained offshore oceanographic measurements of things like temperature, salinity and nutrients, however, ocean scientists slowly figured out that the climate of the Pacific Basin fluctuates over periods of several decades. These cycles, now known as Pacific Decadal Oscillations, or PDO cycles, are characterized by different oceanographic and meteorological conditions. These influence things like ocean currents and water temperature, storm patterns and rainfall distribution, and importantly, the distribution of marine organisms.

Sardines and anchovies alternate in their general abundance with PDO cycles. Cannery Row flourished during a warm PDO period, and when ocean conditions shifted, sardine populations plummeted. Lots of other creatures out there also fluctuate in abundance and distribution with these changing ocean conditions, which have now been labeled regime shifts.

A local history legend, Sandy Lydon, once told me about a semi-professional baseball team named the Monterey Barracudas (who played against the Santa Cruz Sand Crabs a century ago). Sandy, you may not know, knows a fair amount about baseball, and when a bit younger, narrowly missed playing in the big leagues. Well, that's what he told me. My first thought was what a ridiculous name for a baseball team from Monterey; there aren't any barracudas in Monterey Bay. Sandy somewhat politely corrected me, and told me they had been abundant at times, in 1895, 1907-9 and 1915. In looking back at the history of California fisheries, they were even fished commercially during the warm years of 1926, 1931 and 1941, all part of a warm PDO cycle when the sardines also were abundant. Barracuda were also caught in Monterey Bay in 1958, another warm year.

The most recent warm water invader is the Humboldt squid, also known in Mexico as Diablo Rojo or red devil. These are large predatory squids up to six feet long and weighing as much as 100 pounds - not the kind you will probably find on the menu at your favorite restaurant. They are commonly found in the waters of the Humboldt Current off of South America, but have a historic range extending from Tierra del Fuego to southern California. The 1997-98 El Niño event triggered the first recent sighting of these notoriously aggressive predators in Monterey Bay. During the milder 2002 El Niño event the squid returned in larger numbers and have now apparently become permanent residents. They have also been seen in increasing numbers off the coasts of Oregon, Washington and even Alaska. While usually found in deep water, they have been seen closer to shore in shallower water, and have also been known to aggressively attack SCUBA divers who have gotten too close.

Article 10

Weather, Climate, and Coastlines

The sweltering 100+ degree temperatures around Santa Cruz County in early October 2012, might have made global warming believers out of more than a few people. It's important to understand the difference between weather and climate, however. A few very hot days don't make a case for global warming any more than a few freezing days or a heavy winter of snow in the Sierra, argue against it.

Climate is what we predict, and weather is what we get. Climate is long-term. It is about average temperatures and rainfall over large areas and how they change over time. Weather is short-term and local and can vary widely from long-term climate patterns or trends. These differences are often lost on those with particular agendas, however.

We can't just look at any one-day or week or one small area, but need to keep the longer-term picture and trends in mind as we try to determine just what is happening to Earth's climate, if anything.

The slow gradual rise in sea level, which has been going on for thousands of years, and that has begun to accelerate over the past century, is one of those trends we should take notice of.

In 2012 the U.S. ran a distinct fever, with NOAA's National Climate Data Center reporting that the lower 48 states had their warmest January to August period on record. June broke or tied 3,215 high-temperature records across the United States. July was the hottest month ever recorded in the continental U.S. with an average of 77.6 degrees Fahrenheit. By September 2012, 64% of the U.S. was experiencing drought, with August and September comparable to the worst months of the 1930s dust bowl.

Weather in the United States is changing with the rest of the planet, although we may experience temperature extremes and seasonal weather that are slightly different. Since modern global record keeping began in 1880, the 10 warmest years on record have all happened since 1998, with 2010 still ranked as the warmest. 2014 also marked the 38th straight year where global temperature was above the long-term average.

These climatic signals are being recorded along coastlines as well. Tropical Storm Debby in 2012 caused Florida to have its wettest June on record. In August, Hurricane Isaac caused storm surges of up to 15 feet in some places and contributed to Louisiana and Mississippi experiencing their second wettest August on record. Isaac also gave Florida its wettest summer on record.

Many temperature and rainfall records in the U.S. go back a century or more, so when we read that a month was the hottest or wettest on record, and we see this occurring more and more frequently, it's probably a pretty good indication that there is some significant climate change underway.

The warming trend has impacted the Arctic in a significant way as well. Satellite data indicate that the Arctic sea ice extent in September 2012 shrunk to its smallest area since satellite observations began 33 years ago. The minimum ice coverage this year was almost 300,000 square miles smaller than the previous minimum set in September 2007. This is an area almost twice as large as the entire state of California.

As long as the atmosphere continues to warm, sea level will continue to rise because more polar and glacial ice will melt and because the ocean waters expand as they warm. There is absolutely nothing we can do about this over the short term. It's a long-term and complex issue but we need to start making some decisions now.

Over the next several decades, however, sea-level rise most likely won't be our greatest shoreline concern in California. The National Research Council report projects a rise of somewhere between 3 and 9 inches along the California coast by 2030. During a large El Niño event such as 1982-83 or 1997-98, however, sea level can be elevated up to 22 inches for a month or longer. Combined with high tides and large storm waves, it is a repeat of an El Niño like we experienced in one of those winters that will be the most damaging to the coastline in the near term.

However, as sea level continues to rise, and in all likelihood, rise at a more rapid rate, these storm waves and high tides will reach higher and extend farther inland with each additional increment of sea-level rise. This means more frequent coastal flooding and increased bluff and cliff retreat.

Future coastal damages and losses can only be reduced by taking two steps, neither of them easy or immediate: 1] mitigating the warming of the Earth and the resulting rise in sea level by reducing greenhouse gas emissions dramatically, and 2] developing adaptation or response plans for coastal infrastructure and development based on the risks of particular areas.

The first is the most difficult, simply because it involves the entire planet, and because we really only have one atmosphere and one ocean. Progress to date has been minimal at best. The 2nd is a local issue, how will each city, county or region deal with a rising sea? Santa Cruz has now completed both a Climate Change Vulnerability Study and an Adaptation Plan, so if we follow through with these plans, we are on the right course.

Article 11

A Melting Arctic

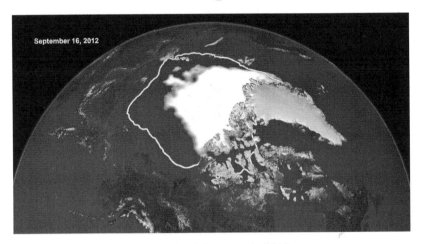

Mininum Arctic ice cover in 2012,
compared to the average minimum in 1979-2010 (delineated by white line).
(courtesy: National Aeronautics and Space Administration)

On September 16, 2012, satellite images revealed that the ice cover in the Arctic Ocean reached its lowest point of the year, and also the lowest since satellites began measuring the amount of ice 33 years earlier. The downward trend in summer loss of Arctic ice continues at about 8% per decade.

There is still a fair amount of ice left in the Arctic, about 3.2 million square miles as of July 2014, give or take a few square miles; but it's melting much faster than predicted and what is left is also getting thinner, so it will melt faster in the future. Is there a problem with this, or might this be a good thing?

Well, it depends. If you are an oil company, the melting of more ice is gradually opening up more of the Arctic for oil exploration. There have been concerns with oil and the Arctic from the time petroleum was first discovered on the North Slope of Alaska in 1968 and a pipeline from Prudhoe Bay to Valdez was proposed. Oil that had to be kept warm enough to flow the 800 miles through the pipeline presented some major engineering problems as the route traversed hundreds of miles of frozen ground.

There was a long list of concerns that emerged as soon as the pipeline proposal surfaced, although most of these were addressed during construction. On balance, because of the visibility of the project, the essentially untouched Alaska wilderness and the potential environmental impacts of oil on wildlife and permafrost, the pipeline was engineered and constructed to deal with the expected problems of transporting hot oil. There were clearly immense impacts of the pipeline construction, including important economic benefits but also significant environmental impacts and social repercussions.

Since completed in 1977, the pipeline has transported about 17 billion barrels of oil. Surprisingly, this amounts to only 28 months of the United States total oil consumption. The pipeline has also experienced several notable incidents of oil leakage, including those caused by maintenance failures, sabotage and even gunshot holes. Due to declining North Slope oil well production, however, it is expected by 2015 that the flow rate in the pipeline will drop to about 500,000 barrels per day, a fourth of its original design.

The British Petroleum blowout in the Gulf of Mexico in April 2010 raised serious doubts about the environmental commitment of the oil industry. Then on July 14 of this past summer, a 570-foot Shell Oil drill ship, part of a fleet heading north for exploratory drilling in a portion of the now ice-free Arctic, slipped its anchor and nearly ended up on the rocks in Alaska's Dutch Harbor. This raised additional concerns in the minds of Alaskans and others about the safety of drilling in the extreme conditions found in the Arctic.

The larger question, however, has to do with the overall global effects of a rapidly shrinking Arctic ice cover. The extent of sea ice in the Arctic grows during the cold dark Arctic winters and retreats when the sun re-appears in the spring and summer. But the sea ice minimum summertime extent, which is normally reached in September, has been decreasing over the last three decades as Arctic Ocean and air temperatures have increased. For several reasons, the Polar Regions feel the impacts of climate change more so than lower latitudes.

The minimum Arctic ice extent in July 2014 was about 714,000 square miles less than the 1981-2010 average for the month. For some comparison, the state of Texas is about 269,000 square miles, so the area of ice loss is about 2.6 times larger than Texas. And some would argue that Texas is really big. If you are having trouble imagining Texas, the ice loss is 4.4 times larger than the area of California.

Sea ice helps keep our planet cool by reflecting most of the sun's energy. Just like you get sunburned when skiing on a clear day due to sunlight reflecting off the snow, the Arctic ice cover also reflects sunlight back into the atmosphere. But as the amount of ice cover is reduced each summer, and the underlying ocean is exposed, that darker water absorbs an increasing amount of heat. This warms the ocean, which then leads to the melting of additional ice. This is known as a positive feedback mechanism.

It's not just the sea ice that's disappearing. The thawing of the permafrost is another problem. Permafrost is soil that is more or less permanently frozen. Nearly one-quarter of the land area in the Northern Hemisphere is underlain by permafrost, including most of Siberia, much of northern Canada and large parts of Alaska.

As the air in the Arctic warms, the permafrost is also beginning to thaw at a faster rate, allowing emissions from the organically rich subsurface sediments to increase. Think of broccoli that you've got in the freezer. As long as it's frozen, it will remain stable for years. But as soon as you pull it out, it gets mushy and soon begins to release the stench of decay. This is exactly what happens when the roots, leaves and other organic matter frozen into the permafrost of the arctic tundra and forests since the last Ice Age begin to thaw.

As the climate warms, the permafrost is thawing, releasing increasing amounts of carbon and methane. The nearly 25% of the Northern Hemisphere underlain by permafrost could potentially release twice as much carbon dioxide as is currently stored in the Earth's atmosphere. And methane traps about 21 times more heat per molecule than carbon dioxide, making methane a potentially much greater concern. Of some minor consolation, the life of a typical methane molecule in the atmosphere is only about 12 years, compared to 50 to 200 years for carbon dioxide.

All of the signs are going in the wrong direction, and to be quite honest, we have wasted enough time debating. Neither carbon dioxide nor methane molecules have political affiliations, they both trap heat. Global climate change is about physical science, not political science, and the science has become increasingly clear.

ARTICLE 12

The Hazards of Arctic Drilling

Gounding of the Kulluk near Kodiak City, 2012.
(photo: Zachary Painter, US Coast Guard via flickr)

The Kulluk, a Royal Dutch Shell drilling barge, was pulled off the coastline of a remote Aleutian island on January 6, 2012, and towed to a protected bay on Kodiak Island where it was to be thoroughly inspected before its next transit. The drilling vessel ran aground on New Year's Eve in the midst of high seas while being towed to Seattle for maintenance and upgrades.

The rig had been drilling exploratory wells in Arctic waters as part of Shell's $5 billion program to rejuvenate its offshore Arctic oil efforts that have been inactive for about 20 years. The melting of increasing amounts of ice in the Arctic has allowed greater access for drilling vessels.

The Kulluk is the second Shell drilling vessel to end up on the Alaskan shoreline in less than six month. It's an odd shaped circular barge, 266 feet in diameter with a 160-foot high derrick in its middle and a funnel-shaped, reinforced steel hull that allows it to operate in ice.

A 360-foot ice-breaking tug, the Aiviq, was towing the drilling barge. On December 27, 2011 during 35-foot seas and 60 miles an hour winds, certainly rough but not unusual Aleutian weather, the main towline to the Kulluk broke. Hours later, while trying to connect to an emergency cable, the huge tug lost power to all four of its engines.

The Coast Guard then sent the cutter Alex Haley to take over towing the drilling vessel, but the rough seas again caused towlines to break. The Haley was forced to return to the station at Kodiak for repairs.

The Coast Guard evacuated the 18-person crew of the Kulluk by helicopter as fears for the vessel's safety mounted. A smaller tugboat, the Alert, was now dispatched to assist in the emergency while the Aiviq's engines were being repaired. The ship was able to again get a line to the Kulluk, but that also failed. The plan was to use the smaller tug to tow the drilling vessel to a harbor on Kodiak, but with the high seas it didn't have enough power and had to disconnect, which is when the Kulluk drifted onto the sand and gravel shoreline.

Finally, the now repaired original tow ship, the icebreaker Aiviq, pulled the drilling vessel off the rocks Sunday night and started the slow tow to a sheltered bay on Kodiak Island. High winds and large swells made progress slow, but by Monday the ships had completed the 45-mile rescue trip. The next step was a complete underwater examination of the ship's hull by inspectors and a sign off by the U.S. Coast Guard before the drilling vessel was towed to Seattle.

The drillship was carrying about 150,000 gallons of diesel fuel and lubricant. This is a very small volume compared to the oil carried by tankers moving along our coast today. For comparison, the Exxon Valdez, which grounded on a reef near Valdez, Alaska, in 1989, about 300 miles northeast of where this recent incident took place, spilled 11 million gallons of crude oil of a total cargo of about 62 million gallons. Fortunately, the shoreline where the Kulluk was stranded was sand and gravel rather than rocks, so no leakage was reported.

While Kodiak Island is a few thousand miles away, and probably doesn't really register here in Central California, on an early Monday morning a few days after the Kulluk grounding, a 752-foot oil tanker, the Overseas Reymar, "grazed" one of the towers supporting the San Francisco Bay Bridge in the fog. Hours earlier, the tanker, with a capacity of about 20 million gallons, had just unloaded its cargo at the Shell refinery in Martinez.

Five years ago, a very similar Bay Bridge collision involving the Cosco Busan, spilled 53,000 gallons of heavy bunker fuel into the bay, fouling 69 miles of shoreline and killing thousands of birds.

Environmentalists have called for a halt to Arctic offshore drilling as a result of a series of problems with Shell's closely watched 2012 effort, including a criminal investigation over issues with safety and pollution control equipment on another of its drill ships. The president of Shell Oil Company said in a written statement about the Kulluk that the company does extensive preparation to ensure such incidents don't occur and is sorry about the grounding.

Shell's foray into drilling off the coast of Alaska has been closely watched, and the latest incident has sparked criticism from environmental groups and some politicians, who have argued that the extreme weather offshore Alaska and in the Arctic makes equipment problems - and hazardous leaks - more likely to occur and more difficult to rectify.

The Alaska Dispatch reported that Shell had attempted to pull the Kulluk through the dangerous weather in an attempt to avoid paying millions in state taxes to Alaska, which it would have owed if the drilling vessel remained in Alaskan waters through January 1. "By risking hazardous conditions to save money, Shell showed a blatant disregard for the safety of the Kulluk crew and the Alaskan environment".

One meteorologist has stated that Shell Oil made a misguided and poorly informed decision to move a huge drilling platform from Dutch Harbor, Alaska to Seattle starting December 21. The Gulf of Alaska is known to be one of the stormiest regions on the planet with one major storm after another during midwinter. The forecasts on the day they left suggested the potential for big storms during the 3-4 week voyage to Seattle.

Then Secretary of Interior Ken Salazar announced a full assessment of Shell's 2012 Arctic drilling to examine "operational issues" associated with the Kulluk, and the drillship Noble Discoverer (which grounded in July), as well as issues with the containment vessel Arctic Challenger and the failure of its containment dome during a Puget Sound test last year. Getting ships into the Arctic and drilling for oil all of a sudden doesn't seem as safe as Shell assured us all it would be.

Congressman Ed Markey of Massachusetts, the top Democrat on the Natural Resources Committee, expressed his concerns in a follow-up statement: "Oil companies keep saying they can conquer the Arctic, but the Arctic keeps disagreeing with the oil companies." After a complete damage assessment of the Kulluk, the repairs were not deemed feasible and Shell decided to scrap the drilling rig in 2014.

PART XI

ENERGY & POWER

Article 1

Oil Formation in the Sea

As gasoline prices edge above $4 a gallon and we realize it's going to cost $50 to fill up the tank, do you ever ask yourself where this stuff is coming from? Most people know that gasoline is derived from oil, which comes from the ground. But there must be some reason why oil is only found in certain places - like Texas and the Middle East - - and not under your backyard.

Oil, believe it or not, is a product of the coastal ocean. Trillions of diatoms and related marine plankton sacrificed their microscopic bodies to power your SUV. Massive blooms of plankton occur in coastal and surface waters of the ocean when conditions are right. We can observe this at times, as we did this last fall, when one particular type of plankton was so abundant that it turned the waters a reddish color (often called a "red tide", although it is unrelated to the tides).

This happens on the Central Coast of California in late spring, and often into summer and fall, when wind moves surface waters offshore, and cooler, nutrient-rich bottom water rises to the surface in a process known as upwelling. The plankton responds like crazy. The process is analogous to turning over the soil and fertilizing your garden in the spring. The plants start growing quickly and soon weeds are everywhere.

These little plankton don't live long, days perhaps. Larger zooplankton or small fish like sardines and anchovies consume many, but massive numbers sink to the sea floor and become part of the accumulating sediment. Their transformation into petroleum involves several steps over a long period of time. First, the organic matter must be covered by sediment or stagnant bottom water to cut it off from oxygen before it can decompose. Next, burial and subsidence beneath additional layers of sediment creates a sea floor pressure cooker, where an increase in pressure and temperature slowly cook the preserved organic matter and convert it into petroleum, or a combination of oil and gas.

These processes have taken place at some unique locations in the world oceans throughout the past 600 million years or so, where all of the right conditions occurred. Some of these oil reservoirs are now preserved on land (Saudi Arabia, Iraq, and West Texas, for example), and others are still under water (the Gulf of Mexico and the Santa Barbara Channel, for example).

Geologists and geophysicists have developed the tools to find these buried deposits, both on land and beneath the sea floor. Extracting the oil is expensive, occasionally dangerous, and it then has to be pumped to shore, transported, refined into gasoline and other by-products, and trucked to your local gas station. It takes a lot to deliver this fossil fuel-solar energy captured and preserved by nature-to your car. And within our lifetimes, what we have now is all we will ever have. Making oil is a process that takes million of years, and we are using it up quickly over a period of decades.

Article 2

Inside the Ocean's Oil-Making Machine

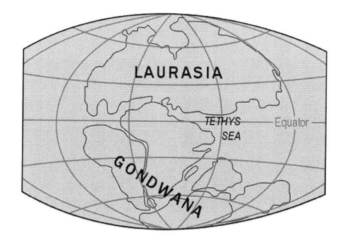

The Tethys Sea in the Triassic period., 200 million years ago.
(© LennyWikipedia, licensed under CC BY 3.0 via Wikipedia Commons)

Oil from the ocean! A few people who read an earlier column were surprised to learn that the creation of oil starts with life in the ocean. But why doesn't oil form in the oceans everywhere? If it did, oil would be easier to find and we'd probably spend a lot less on gasoline. Unfortunately, it's not so simple.

To begin with, the massive blooms of phytoplankton or microscopic floating marine plants required to generate the organic matter that eventually becomes oil, do not occur in abundance everywhere in the oceans. Only in certain geographic areas did it all come together geologically, and those are the places that today hold the world's largest oil deposits. Curiously, five of the seven countries with the largest reserves of petroleum are all in the Persian Gulf: Saudi Arabia, Iran, Iraq, Kuwait and the United Arab Emirates.

Why did so much oil form in the Middle East? To answer this we need to go back about 150 to 200 million years, when the distribution of land and sea was quite different than it is today. Most of the continents appear to have been united into two large land masses, Laurasia and Gondwanaland. The Atlantic Ocean hadn't yet opened up, so North America was connected to Europe, and South America was joined to Africa. The area stretching eastward from this mega-continent, at about the present location of the eastern Mediterranean, was a vast sea known as the Tethys.

This warm body of water straddled the equator and was extremely fertile. There was lots of sunlight, plenty of water and nutrients from river runoff and upwelling. Plankton bloomed in profusion, and for centuries the dead bodies of these tiny creatures sank to the seafloor and collected there. With little oxygen in the overlying bottom water, the organic matter wasn't oxidized or broken down, so it accumulated to great thicknesses as it was covered by sediment. The increased pressure and temperatures at depth over millions of years gradually cooked or converted the organic matter to oil.

About 200 million years ago, as the plates began to break up and the continents shifted. Africa pulled away from South America, India separated from Africa, and both began to move north. As these two large landmasses traveled north toward Asia, they began to compress the Tethys Sea, squeezing and folding the thick layers of sediments on the floor of this ancient ocean with their rich deposits of oil. The waters of the sea drained to the east and west, but the compressed and folded oil-rich sediments became the mountains and deserts of Saudi Arabia, Kuwait, Iraq and Iran. So an accident of plate tectonics made billionaires out of the Saudis and their neighbors, enabling them to trade in their camels for Mercedes.

Although the United States originally had significant oil deposits, what remains today is only a small fraction of what lies beneath the sands of the Middle East. Our appetite for petroleum, however, is huge. With 4 percent of the world population, the United States uses about 25 percent of the oil, or about 20 million barrels every day. About 40 percent of it is imported today (2012), much of it from politically unstable areas. Our imports are down significantly, however, from the 65% we imported two decades ago . Americans used nearly 370 million gallons of gasoline every day in 2013, about 4,300 gallons a second. If lined up in 1-gallon cans along the equator, this daily usage would completely circle the earth nearly six times. No matter how you look at it, this is a lot of oil. And Mother Nature isn't making any more of it, at least within the next several million years.

Article 3

Energy from the Oceans

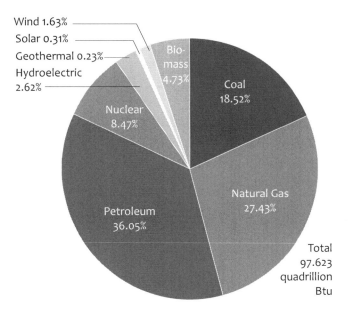

United States energy consumption by energy source, 2013.
(© D Shrestha Ross, with data from the US Energy Information Administration)

Each week I look for a topic with some relevance to current events. Sometimes it works, sometimes it doesn't. Because future energy sources are getting a lot of attention in these pre-election weeks, it seems like a particularly timely topic. While the connection to Our Ocean Backyard may not be immediately obvious, most of our present energy sources, as well as probable future sources, are related one way or another to the ocean.

For now, let's focus on oil, the largest single source of energy for the United States. Oil forms in the ocean, about 27 percent of our current supply is pumped from beneath the seafloor, and much of it is imported and arrives by sea in tankers. With gasoline prices at near record highs, many Americans have voiced support for more offshore drilling.

So let's take a look at the complex issue of U.S. oil reserves and how they stack up against our energy needs.

Before jumping into any discussion of energy and our options, it's important to know what energy sources we rely on today. In 2013, 36% of our U.S. energy was provided by oil, 27% by natural gas, 19% by coal, 8% by nuclear, and 10% by renewable sources (primarily hydroelectric and biomass conversion, but also small amounts of geothermal, wind and solar). Burning fossil fuels that generate carbon dioxide produced 80% of our energy, while only about 10% came from renewable sources.

To put our energy use in perspective, the United States makes up about 4.5% of the total global population, but in 2010 we used 19% of the global energy. As the rest of the world industrializes, standards of living improve, and consumption increases, particularly in China and India, it's probably fair to say that demands on and competition for existing sources of energy will increase substantially for years to come.

In 2011 we used about 4.5 times as much energy per person as China, and 11.5 times as much as each person in India. This is changing rapidly, however.

Much of our energy (27%) is used for transportation: cars, trucks, planes, trains and ships. But comparing our energy demands with those of India and China again, in the USA we have 809 cars for each 1000 persons, almost one for each of us (2011).

China is building cars like crazy, which has led to some horrendous traffic jams and parking problems in places like Beijing. They already have 188 cars/1000 people. India has only 41. But there are more and more people in both nations that are acquiring cars, which means more oil consumption in the future. Bangladesh, in contrast, has 3 cars/1000 people.

Oil is worth looking at more carefully, simply because it remains our most important source of energy, providing just over a third of our total usage. Some feel that this will continue to be our main energy source for the immediate future, and that we should be drilling for more oil. And, of course, the gas that fuels our cars and trucks is refined from oil, and the price per gallon is something that smacks us in the face every time we fill up our tanks. It's also probably fair to say that it isn't ever going to get much cheaper than it is now, as the costs of exploration and recovery from farther offshore, in deeper water, or deeper in the ground, are only going to increase.

We use about 19 million barrels of oil every day in the U.S. (2014), and while we imported 12 million barrels/day a decade ago, or over half of what we consumed, this has been gradually reduced over the past 6 years to about 9 million barrels/day, or about 47% of our use. At the mid-2014 price of about $115 per barrel, this is still about $800 million we pay to foreign countries every day for oil.

Who are we buying this oil from? In 2014, Canada and Mexico provided nearly one half of our daily imports. An additional 24% came from the Middle East: primarily Saudi Arabia, Kuwait, and Iraq (it's likely we would never have invaded Iraq if their biggest export had been broccoli). Brazil, Venezuela, Colombia, Nigeria, Angola, Ecuador, and Russia provide most of the rest. Some of these countries have not been particularly stable in recent decades, nor have they all been fond of the US, which leads some uncertainty in our ability to rely on their oil supplies. We are also increasingly competing with countries like China, whose oil imports have now surpassed ours as the world's largest.

How much oil does this US still have buried underground or beneath the seafloor of our territorial waters? As you might imagine, this is a really complex issue with lots of uncertainties, but I will try to summarize briefly. The most recent estimates (2012) from oil industry research indicate that our "recoverable or proven reserves" (which are defined as the "estimated quantities that are recoverable with reasonable certainty under existing economic and operating conditions") presently stand at 33.4 billion barrels.

This number is a moving target, however. Each year the U.S. Energy Information Agency releases new values, as do various oil companies, based on new exploration, new technologies (fracking, for example), and therefore, new projections.

How long might this oil last? Well, it depends upon how accurate the estimates are, how much we use every day, and what percent of our use we import. You can do the math on your calculator. If you divide 33.4 billion barrels of recoverable oil by our present day total usage of 19 million barrels/day, you get 4.8 years.

That's the worst case, however, if all imports ceased tomorrow and we had to provide all of our own oil. Since we import 9 million barrels per day now, let's assume all of our oil suppliers remain friends for the next few years and we only need to continue to pump the other 10 million barrels/day from our own reservoirs. If you divide our estimated 33.4 billion barrel reserves by 10 million barrels/day, they will only last about 9 years.

If we reduce our oil usage, it will last longer. If we increase our oil use, it will be gone sooner. Either way, the most recent estimates we have on our reserves indicate that we don't have very many years left.

We may find some additional oil offshore, in deeper water, which means that recovery will be more expensive and probably more dangerous. But the oil companies have been exploring for decades and probably have a pretty good idea about where the oil is and how much is out there, so it is unlikely that large new oil fields will be found in the U.S. or offshore waters. In 2008 the federal government offered 160 million acres for lease in the Gulf of Mexico, but the industry only bid on 25% of that area. Why? The most important reason was that the platforms for offshore drilling were very expensive to build; that the existing vessels were already under contract in more lucrative international areas, and that these drill ships couldn't accept any new drilling contracts for the next five years.

Article 4

Energy and the Oceans

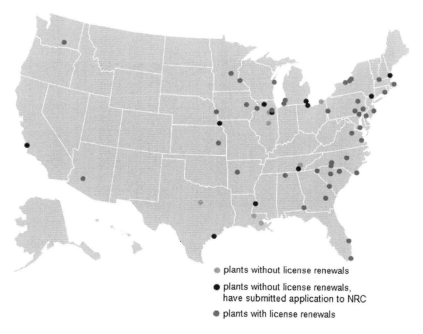

- plants without license renewals
- plants without license renewals, have submitted application to NRC
- plants with license renewals

Location of nuclear power plants in the United States and the status of their licenses, October 2014
(courtesy: US Energy Information Administration)

Important questions about energy confront our newly elected political leaders: What sources of energy will we depend on in the future? How long will each last? What are the impacts of using different energy sources?

In my last column, I wrote that U.S. oil reserves are limited, and even if we decide to increase drilling offshore, it would take at least 5 years to get a platform ready to drill. Well, it's always encouraging to learn that people actually read this column; the president of a

Texas offshore wind power company wrote to correct my statement regarding how long it would take to get a drilling rig ready. "Actually," he said, "if you placed an order today for a drilling vessel, it will take 8 years and a $4 million nonrefundable deposit, on a $200-$500 million investment." Offshore drilling isn't as easy as it sounds, and it isn't an immediate or long-term solution.

There are some potential sources of energy that could come directly from the ocean-wave power, tidal or current power (often referred to as hydrokinetic power or energy), offshore wind power, and ocean thermal energy conversion. But before I discuss those, which are still in early stages of development and installation, I want to cover another major source of energy that has been a part of California's energy history.

Nuclear power had its origins in a December 1953 speech by President Eisenhower on "Atoms for Peace." The first commercial nuclear generator in the U.S. started up in Pennsylvania in 1957. California got the second one, built near Eureka on Humboldt Bay in 1963. In the next decade there was a big push by the power companies to go nuclear in California. Three additional nuclear plants were proposed and constructed (Rancho Seco near Sacramento, Diablo Canyon near San Luis Obispo, and San Onofre in Orange County) and 6 more were proposed. All but one of these were on the coast where they could use the Pacific Ocean for the billions of gallons of water required daily to cool the reactors.

One of these was proposed in 1969 in our own backyard on the coastal terrace just north of Davenport. Some long-time residents may recall the full story, but the upshot was that the proposal was ultimately doomed by a growing popular movement that questioned the safety of nuclear plants and also the unresolved issue of how to safely store or dispose of the radioactive waste generated by the plants. Local communities opposed not only the Davenport plant but also those proposed at Point Arena, Bodega Head (where excavation for the reactor core ran into a branch of the San Andreas Fault and was abandoned), Moss Landing, Malibu, and Long Beach.

Humboldt Bay was shut down in 1976 due to concerns over faulting and earthquakes. Then came the failure of the Three Mile Island nuclear plant in Pennsylvania in 1979 and the explosion of Chernobyl in the Soviet Union in 1986, which heightened public concerns about plant safety. Rancho Seco was shut down in 1989 by a public vote after only 12 years of operation. Only Diablo Canyon and San Onofre are still operating in California (The San Onofre Nuclear Generating Station was permanently shut down in 2013, leaving only a single nuclear plant in California).

In 2014 there were 430 nuclear power plants operating globally, while 115 had been shut down. These plants, however, only provide about 11% of the world's electricity; fossil fuels still provide 67%. Of the 253 nuclear power reactors originally ordered in the U.S. from 1953 to 2008, 48% were cancelled, 11% were prematurely shut down, 14% have experienced at least a one-year-or-more shutdown, and 27% are operating without having a year-plus outage. Thus, only about one fourth of those initially ordered, or about half of those completed, are still operating and have proven relatively reliable.

Today we have 100 operating commercial nuclear plants that provide about 19% of our electrical power. Because of continuing safety concerns, the lack of an acceptable solution for long-term disposal of nuclear waste, as well as the very high cost of plant construction, no new nuclear power plants have been built in the U.S. since 1973. California voters some years ago passed a measure preventing construction of any new nuclear plants in California until a permanent solution to waste disposal was developed. That still has not happened.

In the last several years, however, the nuclear power option has been given new consideration due to growing concerns about fossil fuel consumption and the resulting carbon dioxide emissions and global warming. The Nuclear Regulatory Commission, which licenses nuclear plants, recently reported that 21 power companies say they will seek permission to build 34 new nuclear plants.

There are two primary ocean impacts associated with existing nuclear plants. Both are related to the vast volumes of cooling water used. Diablo Canyon and San Onofre each use about 2.4 billion gallons of water per day. The intake pumps pull in small and large organisms that either die on the filtering screens or, if small enough to pass through, die in their transit through the plant. The discharged water is about 20 degrees warmer than it was when pumped out of the ocean, and the discharge also increases turbidity. The result of all this is a shift in the species that can live in the waters near a plant. These "once-through" cooling systems may not be allowed for any new plants under current EPA rules.

A more fundamental question, however - especially considering Senator McCain's call for building 45 new nuclear plants by 2030 and 100 eventually - is whether we have the uranium resources to power these plants. The best estimates of the global reserves of uranium that can be mined economically is about 5.8 million tons, although the US has only 4% of that. Existing nuclear power plants use 66,000 tons a year (2014). At this rate, the reserves would last about 90 years. However, this rate of usage wouldn't reduce our dependence on fossil fuels or reduce carbon emissions. If we were able to somehow build enough nuclear plants very quickly to replace all fossil fuel usage, our global uranium consumption would be about 1.3 million tons per year. At this rate, our known reserves would last less than 5 years.

An overnight conversion from fossils fuels to nuclear isn't going to happen, but recently the Group of Eight major industrial countries agreed to reduce carbon emissions by 50% by 2050. Using a gradual transition, uranium reserves might last about 10 years. Add in additional uranium that some geologists believe might be available, but more expensive to find and recover, and we might be able to extend this an extra 10-20 years.

Considering the enormous costs ($5 to $7 billion per reactor today) and the large number of nuclear power plants being contemplated to replace fossil fuels, the U.S. could be courting disaster if it chose this route with nothing but blind faith in the continued availability of uranium. There are many good reasons why we need to begin to focus our efforts very seriously on renewable energy options.

Article 5

Energy from the Waves

The wave motor on West Cliff Drive, Santa Cruz, California, c. 1898
(courtesy: Harold J. Van Gorder Collection)

There is something about both oil and nuclear energy that generates strong reactions. My last two columns - which addressed our remaining petroleum and uranium reserves and the time and costs required to develop those energy sources - produced more emails than any others I've written over the past six months.

The objective of my column is to raise awareness and not to champion a particular point of view. But I don't write the headlines, which may sometimes imply a point of view I did not intend.

After the energy crises of the early 1970s had passed, many people lost interest in alternative energy sources. But the recent rise in the price of crude oil and the increased awareness of climate change as a consequence of burning fossil fuels have refocused public attention on the need to develop renewable energy resources. To provide some perspective, renewable sources provided 11% of the energy consumed globally in 2013, and 21% of the energy used to generate electricity. In the US, renewable sources of energy accounted for 10% of energy consumption, and 13% of the energy used to generate electricity. PG&E delivered about 25% of our electricity from renewable sources (17% hydroelectric and 8% wind and solar).

With the exception of nuclear, geothermal and tidal power, almost every other source of energy being considered today, including fossil fuels, has its origin in the sun. Fossil fuels, whether coal, oil or natural gas, are stored hydrocarbons formed initially through photosynthesis, which means they store ancient or fossil solar energy. Hydroelectric power is produced by capturing stream flow, which is the result of precipitation, which depends on evaporation of water heated by the sun. Waves are generated by wind, which ultimately has its origin in atmospheric pressure differences driven primarily by latitudinal differences in solar heating.

We need to look carefully at each of these renewable or sustainable sources to understand the nature and magnitude of the energy available and to consider what may be involved in trying to harness any of these sources. I will also try to summarize the progress that has already been made in developing these sources.

Waves are the most obvious manifestation of ocean energy - ask any surfer, sailor or ship captain about the energy in a large wave. The amount of energy moving through the oceans as waves is impressive.

The energy in a wave is proportional to the square of its height. A wave 2 meters high represents an energy flow of about 25 kilowatts (34 horsepower) for each meter of wave front, enough to light 250 100-watt light bulbs; a wave twice as high will contain 4 times as much energy (2 squared = 4). A single wave 1.2 meters high striking the west coast of the United States may release as much as 50 million horsepower. Calculating how much energy is available is the easier part, but how do we harness any of this energy?

The Armstrong brothers, two local entrepreneurs, actually built what has been called a "wave motor" out on West Cliff Drive near Natural Bridges in 1898. One of the two shafts is still visible on the rock terrace below the bike path just south of Chico Avenue, although it now has a large circular concrete cap over it. There are also several PVC pipes that were placed in the concrete cap to let the compressed air and water out. These blow holes are exciting to watch at times of high tides and large waves.

Although the Armstrong brothers didn't actually generate any power, they did figure out a way to use the energy in the waves to pump water. Taking advantage of a cave and perhaps a natural blowhole, they drilled two large-diameter vertical shafts from the cliff top down into the cave. Two 6-foot-diameter, 600-pound pistons were placed in the shafts, and as the waves surged in at high tide, they pushed the pistons up. As the pistons dropped back down under the force of gravity, they forced water up a pipe and into a water tank at the top of a derrick.

Following the 1897-1898 drought, tourists who came to Santa Cruz were faced with clouds of choking dust as they took their scenic buggy rides along the cliffs. The Armstrong brothers were hired to solve the problem, and with a tank full of salt water high above the cliffs they were able to pipe the water into horse-drawn water tanks that were used to water down the coastal wagon road. Hardly a solution to our energy needs, but an interesting use of wave power nonetheless.

There are now experimental wave energy plants in Japan, Norway, England, Scotland, Sweden, Russia, and India. Portugal had the first commercial wave power project. Although the first phase was short lived and only produced 2.25 megawatts of power-enough to supply 1,500 households-the Portuguese project aims to expand tenfold over the next several years. A milestone was passed in June of 2014, where Phase 2, being tested at the

European Marine Energy Centre (EMEC) off the coast of Scotland, has reached 10,000 hours of grid connected operations. As with all new energy technologies, commercial applications don't happen instantly, and we have to crawl before we can walk.

There are now dozens of research firms working on the technology for harnessing wave power, with names like Pelamis, Anaconda, IWave, Gyrowave, Aegir Dynamo, Dexa Converter, Synchware Power Resonator, Wave Blanket and Limpet. Wave energy is renewable and being an optimist, I believe that it will only be a matter of time until some of these systems become commercially successful and will be hooked up to the grid.

Article 6

Getting Real About Wave Energy

Pelamis, the first system to deliver electricity from wave power to the grid. (courtesy: Pelamis Wave Power)

The waves, tides and currents just offshore contain huge amounts of energy, and this is all renewable. Technologies for converting wave energy into electrical power are evolving, and renewable energy companies are increasingly interested in converting the energy of California's ocean waters into electricity. Nobody is yet doing it, but they have expressed interest.

In 2006, the state legislature passed the California Global Warming Solutions Act, which mandates a timetable for significant reductions in greenhouse gas emissions. Wave energy might help reduce these emissions by providing a renewable and reliable source of energy and could also provide a significant number of new jobs. So what's the hold up?

There are reasons why we aren't all hooked up to an offshore grid quite yet. The principal reason is that the technology is still being developed, but economics, permitting and environmental issues are also standing in the way.

A 183-page report was completed in 2008 for the California Ocean Protection Council entitled "Developing Wave Energy in Coastal California: Potential Socio-Economic and Environmental Effects." While it's dangerous to summarize an exhaustive report in a few sentences, the study concludes that there are not likely going to be dramatic ecological, social or economic impacts of developing wave energy. Nevertheless, it makes a strong case for caution when developing wave energy conversion technology along our coast. At this stage, however, the environmental issues appear manageable. So far, so good.

The development of a renewable ocean energy industry in the U.S. has been hampered to date by a number of regulatory challenges. For example, the regulatory system is not designed to encourage pilot and demonstration projects. State and federal governments need to develop a permitting process that encourages development of demonstration projects while being sensitive to protection of the marine environment. Regulation of ocean power development needs to be clear, efficient and organized with a single lead agency. The European energy community has a test site off the coast of Scotland, the European Marine Energy Centre, set up just for this purpose. We could use one of these offshore California to facilitate testing of new technologies.

A second challenge has been a lack of investment in basic research and development of new technologies. The U.S. Department of the Interior has not funded ocean energy research and development in over 15 years, and it shows. The U.S. doesn't yet generate any energy from the ocean - none. But with appropriations from Congress in 2008, the Department of Energy provided $7.3

million for 14 projects to evaluate the power of the ocean's tides, waves, currents and heat. By 2013, the federal funding for renewable ocean energy development had increased to $16 million for 17 projects. The investment is increasing, but for something with as much energy potential as the coastal ocean, we aren't going to get very far very fast at this rate of investment.

In a report released in October 2008 on the Future of Ocean Power, Greentech Media addressed the underlying fundamentals that will determine when ocean power technologies will become competitive with other renewable and traditional energy sources, what technologies will bring the industry to that point, and how investment, government policies and power industry buy-in will drive the growth of this industry. While fewer than 10 megawatts of ocean power capacity have been installed globally to date (enough to power about 6500 homes), the development of ocean energy has been steadily creeping towards commercial reality each year, with test deployments taking place globally.

The 2008 study analyzed 23 companies that are developing wave energy technologies, of which six have now gone commercial and 4 are listed as pre-commercial.

The technologies that are the most advanced for conversion of wave energy into electrical power fall into four general types:

- An oscillating water column, where the rise and fall of the water in a tube as a wave passes forces air through a turbine that generates electricity.
- Attenuators, such as the snake-like Pelamis, where an elongate and segmented floating tube, about the length of a submarine, flexes as waves pass by, with the up and down motion driving a generator.
- Overtopping, where waves spill over into a floating reservoir with the return flow of water into the ocean driving a turbine.
- Wave buoys or point absorbers, where passing waves cause the structure to rise and fall and uses this motion to turn a turbine.

In September, Portugal completed the installation of the world's first commercial power plant that harnesses wave energy. It uses three articulated steel "sea-snakes" (the Pelamis system) three miles off the country's northern coast. They produced a combined 2.25 megawatts, enough to power 1,500 homes with electricity, While there was optimism for a 10-fold expansion over the next few years, the system shut down not long after going into operation and is now testing the second generation system.

In December 18, 2007, PG&E signed the nation's first commercial power purchase agreement for wave energy, a modest 2 megawatt AquaBuOY project that would capture wave energy off the Humboldt County coast and that expects to begin producing power in 2012. I could find no updates, however, on the success of this facility. The company that is building the buoy, Finerva Renewables, had just received the first license issued for a wave power project in U.S. waters, which would be in Washington State.

The potential amount of energy is high and the technology continues to develop. These efforts need to be encouraged, but we need to streamline the permitting and regulatory process and get some projects in the water. The Electrical Power Research Institute believes that ocean renewable energy in U.S. waters has the potential to provide 10% of today's electrical demand. It is still too early to predict with any certainty how important wave energy will be in our future, but we need to give it a chance.

Article 7

What's Up with Tidal Power?

La Rance tidal power facility along coast of France.
(courtesy: France Energies Marine)

Several columns ago I started to write about energy from the ocean and it has become a far more involved topic than I originally envisioned, and also one that has generated considerable feedback. But, we're not finished with ocean energy quite yet.

Although not widely used, tidal power has the potential for significant future electricity generation. Tides are more predictable than the wind and sunlight, but there is a periodicity to the tides, which limits how many hours each day they can be utilized. Interestingly, using tidal energy isn't a new concept at all. As far back as the 12th century, water wheels driven by the tides were used to power gristmills and sawmills in Europe.

Because the ocean's tides are caused by the gravitational pull of the Moon and Sun on the Earth's oceans, tidal power is essentially inexhaustible. A tidal energy turbine or generator could use the daily or twice daily flow of water into and out of an enclosed bay or estuary, or anywhere else tidal currents are confined, to generate electricity. The greater the water level difference between high and low tides, or the higher the tidal current velocities, the greater the potential for tidal energy generation.

This energy source also produces no harmful by-products, so it's clean. Sounds like a great idea; so why haven't we developed tidal power plants? Well, there are a few.

There are essentially two different ways for harnessing tidal power - either by using the tidal currents themselves, or with a tidal barrier, which would work much like a dam on a river. The first large tidal dam or power station was completed over 40 years ago at St. Milo, along the English Channel coast of France. It has 240 megawatts of capacity, enough for about 190,000 homes (each megawatt or MW of electricity is enough to serve the power needs of about 800 average U.S. homes). A barrier or dam with built-in turbines was constructed across the entrance to the La Rance estuary where the maximum tidal range is 44 feet. That's huge. Electricity is generated on both the flood and ebb tides as the turbines can run in either direction.

A much smaller 20 MW tidal power plant, the first in North America, was built in 1984 on an inlet in the Bay of Fundy in Nova Scotia. However estuaries are highly productive and very sensitive ecosystems and disruption with a dam is a significant concern. Equally important, there are only a limited number of bays or estuaries that are suitable for tidal dams, and the best estimates indicate that using all of these globally would only generate about 1% of the world's electricity needs. South Korea completed a large (250 MW) tidal plant in 2011 and has a second even larger facility under construction. China, Russian and the United Kingdom each have small plants.

The second and more likely approach is using underwater turbines placed in areas with large tidal currents, also known as tidal stream generators. These could be utilized wherever tidal currents are confined or concentrated, beneath the Golden Gate, the Strait of Gibraltar or entrances to large bays or rivers, for example. Since tidal stream generators are still in their infancy, no standard technology has yet emerged as the winner, but a number of different designs are being developed, with some close to large-scale installation. Canada just completed the first such system on southern Vancouver Island in 2006. Modern advances in turbine technology may eventually see large amounts of power generated where natural tidal current flows are concentrated, such as along the coast of Canada, or in the Straits of Florida. Although several prototypes have shown promise with many companies making bold claims, they have not yet been operated commercially or over extended periods of time. But there is always optimism.

Article 8

Ocean Energy from Warm Water and Wind?

Princess Amalia Wind Farm, Netherlands, 2011
(© Ad Meskens, licensed under CC BY-SA 3.0 via Wikipedia Commons)

Ocean energy has received a lot of attention recently, and from the comments posted on-line and emails I've received, it's reassuring to find that people are reading the columns, and to discover their interest and concerns. It's time to move on to other ocean issues but there are two other sources of ocean related energy that I want to cover.

One is Ocean Thermal Energy Conversion (OTEC), a concept that has been around for over a century, but never successfully developed on a commercial scale. It sounds a lot like OPEC, but is clearly worlds apart. The oceans cover 71% of the Earth's surface and are heated daily by the sun, so there is a vast amount of solar energy stored in warm surface waters. The challenge is how to extract it economically. OTEC generates electricity by taking advantage of the temperature difference between the warm surface water and cold water at depth. As in any heat engine, the greatest efficiency is obtained when the temperature differences are the greatest. Differences of at least 20 degrees Celsius (36 degrees Fahrenheit) between surface and deep water are common in tropical latitudes such as Hawaii. However, the relatively small temperature difference means that large flows of both warm and cold water are required.

In a closed OTEC cycle, the warm surface water runs through a heat exchanger to vaporize the water, and the vapor runs through a turbine to generate electricity. The required cold water is brought from the depths up to the surface through the Cold Water Pipe (CWP), and it condenses the vapor back to a liquid. A practical CWP is one of the two key technologies required to commercialize OTEC, and is a huge component, being about 33 feet in diameter and twice as tall as the Empire State Building.

Lockheed-Martin is using modern fiberglass and low-cost composite material manufacturing methods to develop a cost-effective, reliable Cold Water Pipe. Workers at their Palo Alto/Sunnyvale location recently completed an initial small-scale "Proof-of-Principle" demonstration of the new fabrication process, and the company recently received a $1.2 million grant from the U.S. Department of Energy to demonstrate it at large scale. OTEC could enable Hawaii to achieve energy independence, ending its almost total reliance on expensive imported oil.

Wind power has been growing at a pace comparable to solar power with the worldwide capacity increasing at 32%/year, on average, for the past decade. Generation costs have fallen by 50% over the past 15 years and modern wind turbines have improved dramatically in their efficiency and reliability. Over 70 countries around the world are now using wind power, with the U.S. in the lead with 21,000 MW of existing capacity (enough to power nearly 17 million homes), and an additional 8,600 MW under construction. California is 2nd behind Texas in wind energy development with 2,500 MW of existing capacity. A modest tax credit, but one that Congress has repeatedly threatened to eliminate, has spurred the growth of new wind farms; the threat of removing the credit has slowed investment, however.

While we usually think of wind turbines on hillsides, such as Altamont Pass on the way to Stockton, wind farms can also be sited offshore, closer to coastal population centers. Depending upon water depths, different foundation systems are required. One of the largest offshore areas in the U.S. with shallow water is off Cape Cod, where a major wind farm proposal 12 miles offshore, the Cape Wind project is finally moving forward, after years of opposition.

It will be the nation's first offshore wind farm, using 130 large turbines to provide clean, renewable energy for 75% of the population on Cape Cod, Martha's Vineyard and Nantucket. Yet, there has been a decade long drama of opposition. Wealthy homeowners didn't want it to ruin their view. Businesses feared substantial rate increases, and fishermen believed it would interfere with their catches.

While local concerns about wind turbines may have some merit, and need to be considered, they must be balanced against the environmental and social costs of other forms of electrical power generation, such as burning coal.

Wind turbines can also be built farther offshore in deeper water, although foundation costs increase with greater water depth, and costs of connecting to existing power grids also go up. Still, there are advantages to siting wind farms further offshore. Wind speeds tend to be higher and the wind is steadier, which means more energy is available.

A recent study of the offshore wind energy potential along California's coast using existing technologies indicates we have the potential to provide 26 to 112% of the state's total electrical needs, depending upon the height of the wind turbines and wind velocities utilized. This may be our best near-term, renewable ocean energy opportunity.

Article 9

A Drilling Disaster

Wreck of the tanker Torrey Canyon off the coast of England in 1967.
(photo: Jackie Hamilton, courtesy: the Living Memory Association)

Unfortunately, there is absolutely nothing positive to say about the April 2010 explosion and well blowout in the Gulf of Mexico. It's a tragedy for everyone and everything involved. The waste of oil, the loss of human life, the impacts of the oil on the ocean and coastal environments, and the economic losses will be with us for years.

Most of the initial news coverage was focused on how much oil actually was released into the ocean (with the total amount ultimately determined to be nearly five million barrels or 210 million gallons), the failure of each of the methods used to retain or capture the oil, speculation about how long and what it might take to seal off the well or somehow reduce the release of oil, and the concern for the biological losses expected along the Gulf Coast and beyond. The blown out well gushed oil for 87 days before finally being capped.

The 1989 Exxon Valdez spill near Valdez, Alaska, released 11 million gallons of oil but the remainder of the cargo was contained in a tanker floating at the surface under relatively calm conditions. In the case of the British Petroleum's Deepwater Horizon drilling platform, the oil was released from a pipe broken in three places on the seafloor, 40 miles offshore at a water depth of 5000 feet. Not exactly like fixing your backyard garden hose. The broken pipe penetrates an oil reservoir under high pressure containing an undisclosed amount of oil, but certainly millions of barrels. In all likelihood, this will be the worst oil spill in U.S. history.

In my very first Ocean Backyard column in April 2008, I explained how petroleum forms in the ocean. Several months later I talked about our use of oil and where it comes from. As a quick summary, while the U.S. has 4.5% of the world's population, we use 25% of the oil, about 20 million barrels every day. When I first wrote this column in 2010, we were importing about 2/3 of this or about 11 million barrels every day. At the going price of crude oil of $80/barrel at that time, we were spending $880 million a day buying foreign oil. By April 2014, our imports had dropped to 9 million gallons/day but price had risen to $95/barrel, so we were still paying out about $855 million to buy foreign oil every day.

We are fortunate to have two next-door neighbors, Canada and Mexico, who can be counted on for nearly half of our imports. But much of the rest comes from places like Angola, Nigeria, Kuwait, Saudi Arabia, Iraq, Venezuela and Russia. While our imports have gone down over what the last six years, we are still spending nearly a billion dollars a day importing oil, with much of the money going to countries whose interests don't necessarily align with ours.

So "drill, baby, drill" is one of the ways that was proposed during the 2008 election to help alleviate this situation. One major limitation I described at that time was the shortage of offshore drilling platforms. I was surprised to hear from an industry executive who evidently read the column and wrote then to let me know that if you placed an order for a drilling vessel at that time, it would have taken 8 years, and a $4 million nonrefundable deposit on a $200-$500 million investment. This is probably one reason why BP told the government during their blowout that it might be at least three months until they could get a rig on the site, with the hope of drilling to reduce reservoir pressure. One thing that does seem clear is that there wasn't a plan in place to deal with a blowout in 5000 feet of water. If it happened again tomorrow, would we be in any better shape?

Article 10

Drilling and Spilling

There wasn't much good news in the months immediately following the 2010 Gulf of Mexico well blow out. The three companies involved, British Petroleum, Halliburton, and Transocean mostly argued about who was to blame. The use of booms to contain the oil was ineffective, and the initial efforts to cap the well were also unsuccessful.

The initial BP report that only about 5000 barrels of oil were being released each day by the leaking well seemed like an underestimate to Richard Harris, a National Public Radio science reporter. He obtained a video taken by a submersible of the leaking wellhead and asked three different experts to independently try to calculate the actual volume using their own distinct expertise. A Flow Rate Technical Group was also charged with making the same calculation. Their determinations using several different methods indicated that the actual leakage was 5 to 10 times larger than BP initial estimates.

One of the first very large oil spills was the wreck of the tanker Torrey Canyon off the coast of southwest England in 1967. This huge ship went aground on a rock reef in broad daylight as the shipmaster was taking a short cut to save time. This accident brought two issues to light: the problems of spill responsibility and the difficulties of cleaning up an oil spill in the ocean. The ship was owned by Union Oil, was chartered by British Petroleum, registered in Liberia, carrying Kuwait oil to England, and had a German captain and an Italian crew. Whose fault was it? The 30,000,000-gallon spill ultimately covered over 700 square miles of sea surface, and fouled 120 miles of English coastline and 50 miles of French coast.

The Royal Air Force, with virtually no experience in such matters, dropped forty-two 1000-pound bombs, cans of aviation fuel and napalm on the slick in an unsuccessful effort to get it to burn. 10,000 tons of very strong detergents were also used to break up the spill but were generally believed to have done more biological damage than the oil itself.

Two years later in 1969, a well being drilled by Union Oil from a platform 6 miles off Santa Barbara in 275 feet of water blew out. If you must have an oil spill, the coast of Santa Barbara isn't the place to do it. Over the next 11 days about 3,000,000 gallons of oil formed an 800 square mile slick that took a large toll on marine life, particularly sea birds, as it coated the coastline of Santa Barbara with a black sticky goo. Many viewed this disaster as the event that gave rise to the modern environmental movement.

The well continued to leak for months. Like the Torrey Canyon spill, clean up efforts were again primitive and problematic. Straw was used in attempt to soak up oil on the beaches, detergents were sprayed in the water, and steam cleaning and sand blasting of the intertidal zone took place, with an additional toll on marine life. No matter how you look at it, cleaning up spilled oil is messy, ugly and leaves no one satisfied or happy.

Twenty years later we had the Exxon Valdez spill in Alaska, and clean up methods had still not advanced much further. Many scientists now believe that any efforts to clean up the tarry residue on the coastline are probably more damaging to marine life than the oil itself as they only remove the surface coating at best. In the Gulf of Mexico, several weeks after the initial explosion, the engineers were talking about a "junk shot", trying to plug the well with golf balls and rope! Is this really the best we can do in this industry?

Somewhat surprisingly, however, these highly visible spills, whether drilling accidents or tanker or pipeline spills, on average constitute only about 11% of the 30 million gallons of petroleum that enters North American waters each year. Nearly 85% comes from land-based runoff, polluted rivers, small boats and jet skis.

Article 11

Shuttering San Onofre

San Onofre Nuclear Power Plant.
(© 2002 Kenneth & Gabrielle Adelman,
California Coastal Records Project, www.californiacoastline.org)

On June 7th, 2013, Southern California Edison announced that they were going to permanently close the San Onofre Nuclear Generating Station (often referred to by the acronym SONGS, although there is nothing remotely musical about a nuclear power plant), perched on a coastal bluff in northern San Diego County.

At full capacity, San Onofre provided power for about 1.5 million homes in Southern California, however the plant has had a number of mechanical problems since it first went on line in 1968. Unit 1, a first generation reactor, operated for 25 years and closed permanently in 1992. Units 2 and 3 went into commercial operation in 1983 and 1984, but had received multiple citations over the years. These two reactors have been shut down since January 2012 due to premature wear found on tubes in the steam generators, which apparently contributed to the accidental release of a small amount of radioactive steam.

That release, coming not long after the meltdown of the Fukushima Daiichi nuclear plant in Japan in early 2011, set off a flood of public opposition and a regulatory and legal struggle between the Nuclear Regulatory Commission (NRC), Southern California Edison and Mitsubishi Heavy Industries, which built the flawed parts.

The closing and decommissioning of San Onofre, which will cost in excess of $4 billion, paid for primarily by the customers, leaves one remaining operating commercial nuclear power plant in California. Diablo Canyon, on the coast south of Morro Bay, has struggled with its own issues over the years, most recently the newly discovered Shoreline Fault, which lies a short distance offshore. The recent detailed mapping of this near shore fault by the U.S. Geological Survey provides new information on the length of this active fault and its potential for generating earthquakes.

Forty-four years ago, on April 9, 1970, Pacific Gas & Electric Company held a press conference in Santa Cruz, announcing that they had taken out a lease on 6800 acres of coastal land north of Santa Cruz owned by Coast Dairies and Land. Their proposal was to build what would have been the world's largest nuclear power facility on the coastal terrace just north of Davenport.

I was in that room 44 years ago and with a handful of others, immediately began to wonder what this might mean for Santa Cruz. Why here? Santa Cruz was a small, somewhat isolated community at that time and certainly didn't need all that power.

Well, there were several reasons. Most importantly, the ideal place to build a large power plant of any type is close to the center of the demand, in this case, the rapidly suburbanizing Santa Clara Valley. And with a thermal generating station, whether fossil fuel, like Moss Landing, or nuclear, you want to be next to a large source of cooling water, the Pacific Ocean. There was also an objective of placing plants, particularly nuclear plants, still somewhat new at the time, away from large population centers. Davenport was seen by PG&E as the perfect location.

In the 1960s and 1970s, everything was different in California. Growth and associated electrical demand were increasing rapidly. The Resources Agency in a 1970 report maintained that a growing population and increasing per capita consumption of electrical power would double the electric power demand in California every nine years. Nuclear power was seen as an answer, and was advertised at that time by PG&E as "efficient, economical, safe and clean".

The state has had a long and checkered history with nuclear power plants. There were two that were built and shut down years ago, Humboldt Bay and Rancho Seco, near Sacramento. Although not generally known today, there were also others proposed at Point Arena, Bodega Head, Moss Landing, Malibu and Long Beach. These were all proposed along the coast for access to large volumes of cold seawater. None of these were ever built, although approval was received for Bodega Head and excavation had started on the foundation when a branch of the San Andreas Fault was discovered in the pit. The seismic problems on the site as well as public opposition led to termination of the Bodega Head project in 1964.

Article 12

A Nuclear Power Plant for Santa Cruz

The April 9, 1970, Pacific Gas & Electric Company announcement that they planned to build a large nuclear power plant on the coastal terrace just north of Davenport set off a community debate that went on for several years.

On one side, there were those who believed that the creation of new jobs and increased tax base would be a boon for Santa Cruz, whose economy was still based in large part on summer tourism. On the other side was an embryonic environmental community concerned about the potential environmental impacts and safety of a nuclear plant. At that time there were only a handful of nuclear plants across the country.

PG&E, in their advertisements at the time, stated rather emphatically that nuclear power plants were efficient, economical, safe and clean. They were being perceived by some as the wave of the future, like all electric homes.

A local group soon formed with the acronym of CEDAR, the Committee to Examine the Dangers of Atomic Reactors. This citizen's group, as well as the first Environmental Studies class on the UCSC campus, the Environmental Workshop, both began to look carefully at these four claims.

The class of about thirty students produced a booklet titled Santa Cruz and the Environment, which took a lot of local heat for publicizing the environmental issues affecting Santa Cruz at that time, including the proposed nuclear plant. This was a striking contrast to the many environmental groups today that are actively engaged locally in every environmental issue that arises, large or small, local or global.

An investigation of the advertised efficient, economical, safe and clean nature of nuclear plants revealed that none of these claims were really true.

Although nuclear power represents a highly concentrated source of energy, it doesn't achieve the high temperatures of burning oil or coal, so the steam produced doesn't have the necessary temperatures and pressures for efficient conversion of heat to electricity. As a result, a nuclear plant operates at a lower overall efficiency and requires considerably more cooling water than a fossil-fuel plant of the same capacity.

A closer look also revealed that nuclear plants were far more expensive to build than conventional plants. Much of the cost of the early plants was covered by subsidies, including about one-third of the construction costs coming from the Atomic Energy Commission, which had the conflicting roles at that time of both promoting the use of nuclear energy but also regulating it. The federal government was also subsidizing the insurance, the fuel, as well as the removal and storage of radioactive waste. These subsidies reduced the costs to the power companies significantly, but were all borne by the federal taxpayers.

Safety is a much more serious issue with a nuclear plant than a fossil fuel plant, simply because the accidental release of radiation can be carried over large distances by wind and water, stays around for a long time, and can produce both immediate and long-term impacts. While there had been a number of accidents in test and experimental reactors at that time, there had been relatively few commercial plants in operation such that their overall safety record was good.

The partial meltdown of the Three-Mile Island plant in Pennsylvania in 1979, followed by the Chernobyl disaster in the former Soviet Union in 1986 with the release of radiation that spread over much of western Russia and Europe, and then the Fukishima Daiichi disaster in 2011, all contributed to a much greater public concern for the safety of nuclear plants. Ultimately nuclear plants all depend upon error free construction and operation, and unfortunately humans are not quite perfect.

While nuclear plants don't produce the visible emissions of fossil fuel plants, they do produce radioactive wastes. Of greatest continuing concern has been the spent fuel, high-level radioactive waste that must be isolated from humans and the biological environment for thousands of years, which in the United States is still being stored in huge tanks in the states of Washington, Idaho and South Carolina. No permanent solution has yet been agreed upon such that California banned any new nuclear plants decades ago until this waste storage/disposal problem had been resolved.

Ten nuclear power plants were proposed over the years along the California's coastline; only four were ever built; three of those have now been closed.

PART XII

WATER
FRESH & SALTY

Article 1

Why the Oceans are Salty

The city of Santa Cruz Water Department and the Soquel Creek Water District worked with the University in 2008 to construct a pilot desalting facility at Long Marine Laboratory. The lab had a seawater intake directly from the ocean so there was an easily accessible supply of salt water. This experimental plant was constructed to try out various filters and membranes in order to determine if desalination was a viable option for the seawater along our coastline.

The actual plant being considered jointly by the two water districts was one of several proposed along California's coast in recent years to augment our fresh water resources through the process of desalination. Existing supplies are insufficient to meet our needs during low rainfall periods or drought periods and there is an expectation that as climate continues to change, that droughts may become more frequent.

Why are the oceans filled with salt water anyway? Wouldn't it be a lot easier, as well as a lot less expensive, if the oceans were fresh and all we had to do was just pump the water out?

What may be surprising is that 97% of all of the water on Earth is in the oceans, is salty, and not terribly useful to us for most human uses. Of the remaining 3%, almost 2% is pretty much out of our reach because its frozen as ice sheets and glaciers. This leaves about 1% of all the water on Earth for us to fight over, which is what we have done in California and a lot of other arid or semi-arid places. In fact, Mark Twain is credited (many say incorrectly) with writing over 100 years ago that in California, "whiskey's for drinking and water's for fighting over".

How salty is the ocean anyway? Well, too salty to drink is the short answer. The total amount of salt in seawater is only about 3.5% by weight; so a gallon of seawater only contains about 4.5 ounces of salt. That's not much, but its enough to make it totally undrinkable and of virtually no use for agriculture. The most common elements or ions that make the seas salty are sodium and chloride (which when combined make table salt), but there is also a long list of other dissolved elements or ions in seawater. While there are only 4.5 ounces of salt in each gallon, if we took all of the salt out of the world oceans it would be enough to cover the entire planet with a layer about 500 feet thick!

So why is the ocean salty anyway? What may be a little surprising is that river runoff is one of the primary sources of the salt in the oceans. While we think of the water in our streams as "fresh", our creeks and rivers actually contain small amounts of dissolved material from the erosion and weathering of the rocks in their watersheds. This dissolved material or salt has been accumulating in the oceans for nearly four billion years. Some dissolved material is also added from thermal vents or hot springs on the seafloor, and some from volcanic eruptions. Without a way to remove salts, however, seawater would continue to get saltier. We know from analyzing fossil marine organisms and also salt water preserved in some sedimentary rocks, however, that the salt content or salinity of the oceans hasn't changed significantly over geologic time.

There must be some processes or mechanisms by which the salt in the ocean is removed at the same rate it's added in order to keep the salt content in balance. What may at first glance seem like an unimportant process is the salt spray that is created as waves break along the shoreline. You can feel the salt on your skin walking along the cliffs on a day with crashing waves when lots of spray is in the air. This process removes an enormous amount of salt. Water also seeps or infiltrates into the ocean bottom near hydrothermal vents on the seafloor, removing dissolved material. Large volumes of salt are also precipitated in salt flats or tidal embayments, such as San Francisco Bay, where seawater evaporates and leaves the salt behind. Additional ions or elements are used by marine organisms to make their shells, while still other elements are attracted to the surfaces of clays and biological particles as they settle to the seafloor to be incorporated into sediments.

The salinity or salt content of the oceans is in balance, and has remained more or less constant for hundreds of millions of years. Life in the sea has evolved to live comfortably surrounded by salt water. Desalting seawater, which today may seem like an exotic and expensive way to produce fresh water, may likely become increasingly more competitive in cost with other sources of fresh water in the future, particularly if global climate change projections lead to a reduction in the availability of fresh water, which seems to be the trend we are witnessing. While we can live without oil and gas, and if healthy, may live for up to three to four weeks without food, we can only survive about three or four days without water.

Article 2

Five Years and 130 Columns

In May 2013, I realized that it had been five years since I wrote my first Ocean Backyard column, thinking at the time that it might last six months or so, and then I'd probably run out of things to write about. Looking back at this column in August 2014 its now been 165 columns and the topics seem to still keep jumping out at me from the keyboard, from NPR, from talks with friends, or emails from readers.

Once I get started writing an article, I've discovered that it's easy to get carried away. The editor has asked me more than once to shorten my columns. While this is good discipline, it's also difficult at times to contain my word flow. "If I had more time, I would have written a shorter letter", is a quote that has been attributed at various times to a number of famous individuals, including Benjamin Franklin. So I try to obey the editor. It just takes a little longer to write a shorter column.

As a way to mark the first five years of articles, I decided to do something different and write a more personal column. I realize when I read an editorial or column, I've often wondered who is this person?

Well, in contrast to many scientists, who are quite comfortable within their own discipline, talking and exchanging ideas with colleagues who study similar things, I've found communicating with non-scientists or the public to be just as interesting, and perhaps even more challenging. So I do a lot of that.

This is my 46th year of teaching. I was made aware of how long it had been about 15 years ago when a young student came up after my first Oceanography lecture of the quarter to inform me that her mother had taken my class. It's happened a number of times since then and I've come to see it as a badge of some honor.

My ocean interest really grew out of two things: 1) a lot of time spent outside as a kid, camping for a good portion of every summer, and living on a farm in Oregon for several years, which led me to develop an interest in the natural world; and 2) surfing, which drew me more to the ocean part of the planet.

I was quite impressionable as an undergraduate student at UCSB in the 1960s, changing my major five times before graduating in geological sciences. A poster of oceanographers working off the side of a ship in the North Pacific and some encouragement from my professors led me to apply to the graduate program in Oceanography at Oregon State University. I wrote letters to a number of other oceanography graduate schools, much better known at the time, like Scripps and Woods Hole, but I was a bit broke at this stage in my educational career and I opted for the school with the lowest application fee - not usually a good idea.

Fortunately, it turned out to be a good decision. I spent a lot of time at sea and found the oceans to be even more fascinating than I had imagined. I managed to finish my Ph.D. degree in three years, and was within a few days of accepting a job in Houston, working for Exxon in offshore oil exploration, when I got a call from the newly opened University of California campus at Santa Cruz. The Earth Sciences Department was looking to hire the university's first oceanographer. I said yes and my entire career and life changed. Yes, it definitely was a good decision and I have never for a moment regretted that decision.

ARTICLE 3

A Primitive Ocean

In the beginning, there was no ocean backyard, and in fact, no ocean. When the Earth came into existence about 4.6 billion years ago, several processes conspired to heat up the primordial planet and thereby prevent the accumulation of liquid water: bombardment by asteroids, comets and other debris from space, all warmed things up; gravitational compression of accumulating debris within the Earth was converted to heat; and decaying radioactive materials also raised the planet's temperature.

Over the next 500 million years or so, as the early Earth slowly cooled, water began to collect in the depressions and low areas. To date, the oldest sedimentary rocks found on Earth are about 3.9 billion years old. Because sedimentary rock such as sandstone, shale or limestone require water in order to form, there is good evidence that seas or oceans have been around for nearly four billion years.

Scientists have asked for centuries, where did all this water come from? There are two likely sources, but no complete agreement on the importance of each.

The favored idea for decades was that the water at the Earth's surface, in the atmosphere, and oceans, came from within the Earth. Magmas, which erupt from volcanoes on the surface, typically contain a few percent by weight of dissolved gases. These gases include carbon dioxide, nitrogen, sulfur dioxide, and water vapor, which is typically the most abundant.

Measurements of gases taken around volcanoes during eruptions, always a hazardous endeavor, have given us some reasonable estimates of the importance of this source of moisture over geologic time. Given the approximate amount of lava erupted by volcanoes over the history of the planet, and some average water vapor contents, the amount of water produced by volcanoes is equivalent to roughly one hundred times the volume of the world's oceans. In other words, there is no problem squeezing the world oceans out of the water contained within the Earth, all 325,000,000 cubic miles of it.

The importance of this more or less accepted source of the water in the sea has been questioned recently, however, with a less obvious source put forward - large volumes of water likely have been added over time from space. Icy comets or meteorites from the far reaches of the solar system colliding with Earth throughout its history may well have also contributed significant volumes of water that helped cool the Earth's surface and gradually collected to help form the oceans.

As the primordial oceans began to form, their water began to dissolve minerals in the rocks at the Earth's surface. Large volumes of water were also being evaporated from the oceans, which then condensed in the atmosphere and fell back down as rain. This precipitation and runoff through the earliest rivers and streams flowing towards the low areas or oceans, also helped to break down the rocks and transport their chemical constituents to the sea. In addition, many of the ions that contribute to the salinity of the oceans are believed to have come from within the Earth, from both volcanic eruptions and also thermal vents on the sea floor, over millions of years. As a result, the oceans began to accumulate these ions and become salty very early in their history.

ARTICLE 4

Salt in the Sea

The oceans began to accumulate salt as soon as they started to form nearly four billion years ago. The ions of sodium, chlorine, magnesium, calcium, and just about every other element you can remember from the periodic table in your high school chemistry class, came from several different places. What may be a little surprising is that river runoff is one of the primary sources of the salt in the sea.

While we think of the water in our streams as "fresh", our creeks and rivers, whether the Colorado, the Amazon, or the San Lorenzo, actually contain small amounts of dissolved material from the erosion and weathering of rocks in their watersheds. This dissolved material or salt has been accumulating in the oceans for billions of years. The amount carried by a stream varies depending upon the geology of the drainage basin, as well as the rainfall and climate of the region. Dissolved material is also added from thermal vents on the seafloor, and some even comes from volcanic eruptions.

Without a way to remove salts, however, seawater would continue to get saltier as more dissolved minerals enter the ocean. We know from analyzing fossils of ancient marine organisms and also from the composition of the water sometimes trapped within older marine sediments, that the salt content or salinity of the oceans hasn't changed significantly over geologic time. Water wells drilled into some sedimentary rocks in the Santa Cruz Mountains actually have pumped up ancient seawater, with a salt content essentially the same as the present day ocean.

So there must be some processes by which the salt in the ocean is removed at the same rate it's added in order to keep the salinity constant. Some elements, calcium and silica, for example, are extracted by marine organisms to make their shells, Other elements are attracted to the surfaces of clays or biological particles as they settle through the ocean to the seafloor where they incorporated into sediments.

What may at first glance seem like an unimportant process is the salt spray that is created as large waves break along the shoreline. This salty air can take its toll on anything that is resident along the coast: a car, a bike, even the paint on a house. You can feel the salt on your skin walking along the cliffs on a day with crashing waves. This process can remove a large amount of salt.

Water also seeps or infiltrates into the ocean bottom near hydrothermal vents on the seafloor, removing dissolved material. Large volumes of salt are also precipitated in salt flats, on the edges of tidal embayments or lagoons, such as the margins of San Francisco Bay, where seawater evaporates and leaves the salt behind.

How salty is the ocean anyway? Well, too salty to drink is the short answer. The total amount of salt in seawater is only about 3.5% by weight. A gallon of ocean water contains about 4.5 ounces of salt. That's not much, but its enough to make it totally undrinkable and of virtually no use for agriculture. The most common elements or ions that make the seas salty are sodium and chloride (which when combined make table salt), but there is also a long list of other dissolved materials in seawater. While there are only 4.5 ounces of salt in each gallon, if we took all of the salt out of the world oceans it would be enough to cover the entire planet with a layer about 500 feet thick!

Article 5

Salt from the Sea
Part 1

Unfortunately for us, every gallon of seawater contains about 4.5 ounces of salt. It doesn't sound like much, but it's enough to make it essentially undrinkable and pretty toxic for most crops; well, unless you're cultivating seaweed. And there are somewhere around 325 million cubic miles of it out there, covering nearly three-fourths of the Earth's surface. While coastal waters are relatively shallow, on average the oceans are over two miles deep. This vastness became abundantly clear to me on our recent round the world voyage when it took us almost 18 days just to cross the Pacific. There is a just a lot of salty water out there.

Although the average salt content is about 3.5%, the salinity of the oceans varies a bit from place to place, particularly in coastal areas. It can be essentially fresh a number of miles offshore where large rivers like the Amazon discharge, and saltier than normal where evaporation is very high, places like the Red Sea or tropical lagoons.

No matter where we are in the world oceans, however, and no matter what the specific salinity is at that particular location, the % of the salt contributed by each of the individual ions or elements remains exactly the same. Stay with me here. Chloride always makes up 55.1%, Sodium 30.6%, Sulfate 7.7%, Magnesium 3.7%, Calcium 1.2%, Potassium 1.1%, and on through the periodic table. These six most abundant elements or ions, however, make up 99.36% of the total salt content. The ocean is really like a massive slow motion blender, with currents constantly mixing up all the salts or ions that are delivered by rivers, volcanoes, and seafloor vents and distributing these individual elements evenly throughout the world oceans.

This raises an interesting issue, in this era of trying to be natural and organic: No matter what label or name is on that container of salt you buy (sea salt, organic salt, extra-virgin salt, etc.), if it comes from evaporated seawater, its all essentially the same stuff. Well, except many salts have iodine added, which is an important trace element.

All the world's salt comes from the evaporation of seawater, whether this took place millions of years ago from ancient oceans where it was preserved as rock salt in places like Kansas, New Mexico, Texas, Florida, and a number of other states; or it is harvested from evaporation ponds along the edges of today's oceans, bays or lagoons. South San Francisco Bay, for example, has had evaporation ponds for salt production operated by Cargill for decades. If you want to see some amazing images of these ponds, Google "San Francisco Bay salt ponds" and you will see what look like ponds with bright food coloring; but in fact the colors are from different types of algae that grow in the ponds under different salinity conditions.

There were also salt ponds on the north side of Elkhorn Slough for years, and some of the old buildings are still visible just on the north side of the harbor bridge. The Salinas River was named from the world salina, meaning salt flats or salt works, and the river formerly flowed through this area before discharging several miles north of the present day harbor entrance. This area on the old 1910 and 1933 maps is clearly labeled "Monterey Salt Works" and "Salt Farm".

Part XII: Water: Fresh & Salty

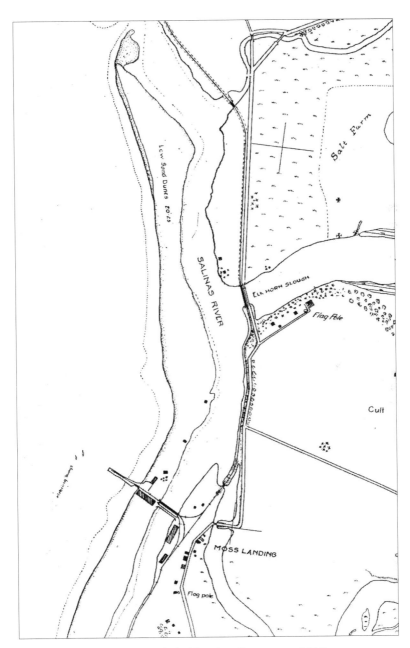

Map of Saltworks in Moss Landing area, c. 1910.
(courtesy: US Coast and Geodetic Survey)

Article 6

Salt from the Sea
Part 2

Former Kaiser Refractories, Moss Landing, California.
(© 2010 Gary Griggs)

In this time of extreme drought in California, we don't really have a water shortage, we just have a fresh water shortage. Unfortunately for us, nearly 97% of all of the water on Earth is salty and is in the oceans where it is not particularly useful to people. Once separated from sea water, however, salt, or more precisely, halite or table salt, has always been a valued commodity. Saltiness is one of the most basic human tastes, and as a result, salt has long been one of the most widely used seasonings. It has also been used for preserving food for centuries.

Virtually all of the earliest civilizations harvested, traded or used salt; whether the Chinese, Hebrews and Hittites, Greeks and Romans, or Byzantines, they all valued salt. The earliest evidence of salt ponds or extraction date back about 8000 years. Salt was traded across the Mediterranean by ships, carried across the desert by camels, and even used as currency in parts of Africa.

Those peoples or regions that had salt deposits grew wealthy from the trade or use of salt, and many governments throughout history have taxed salt as a way of generating income and subjugating people. It is believed that funds generated by salt production in southern Spain financed the voyages of Christopher Columbus, and the salt tax in France was one of the causes of the French Revolution.

Interestingly, the word "salary" came from the Latin word "salarium", which referred to money paid to the Roman soldiers to purchase salt. Even more bizarre, the word salad, literally means "salted", and originated from the ancient Roman practice of salting otherwise bland leafy vegetables.

In India in 1930, Mahatma Gandhi led a 24-day, 240-mile Salt March (usually known as the Salt Satyagraha) as a nonviolent protest of the British salt monopoly and tax in colonial India. This march drew worldwide attention to the Indian independence movement. Along the march, Gandhi and his followers made salt from seawater.

About 30% of all of the world's table salt used today is extracted directly from seawater, usually from solar evaporation in coastal salt ponds. This is most effective in hot dry climates where evaporation rates are high. Evaporation allows the water to move into the atmosphere in a vapor phase while all of the dissolved salts are left behind, ultimately producing concentrated brine that contains several different "salts". Through a refining process, sodium chloride or table salt is separated from the other salts.

At the other extreme, where the climate is cold, salt can be extracted by the freezing of seawater in coastal ponds. Just as salt doesn't evaporate with the water, when ice forms from seawater, the salt ions don't fit comfortably in ice crystals and are squeezed out and concentrated beneath the ice as a salty brine.

I think most of us probably think of table salt as the primary use of the salt in the sea, but in fact a number of other elements are extracted from ocean water as well. About 70% of the global supply of bromine comes directly from seawater, as does 60% of the magnesium.

Magnesium was extracted from the waters of Monterey Bay for decades by Kaiser Refractories, which operated the industrial complex immediately south of the Moss Landing power plant. These are the large tanks that you can see from Highway One driving through Moss Landing. The magnesium was used during World War II for making bombs, and later magnesium oxide was used for manufacturing high temperature bricks for use in steel furnaces. Imports of lower cost magnesium from China led to the closure of the Kaiser plant some years ago, however.

Article 7

Fresh Water

A Drop in the Bucket

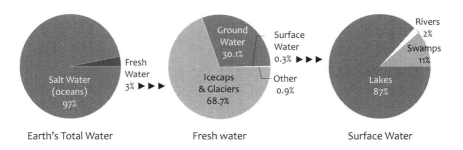

Distribution of water on Earth
(© D Shrestha Ross)

What may be surprising and not particularly good news, is that nearly 97% of all of the water on Earth is in the oceans, is salty, and not terribly useful to us. Of the remaining 3%, about two-thirds is frozen as ice sheets, glaciers, and permafrost, much of this out of our reach in Antarctica and Greenland. Although there were proposals decades ago to wrap large icebergs from Antarctica in plastic wrap and tow them to the thirsty people, lawns and golf courses of southern California, I think there is about as much chance of that happening today as there is of Santa Cruz getting 6 inches of rain in August.

Unfortunately, this only leaves about 1% of all the water on Earth for us to fight over, which is what we have done in California and a lot of other arid or semi-arid regions across the west. Mark Twain is often quoted as saying over 100 years ago that in

California, "Whiskey's for drinking, water's for fighting over." There is a long-standing argument, however, about whether Mark Twain really made this statement. Whether he did or didn't doesn't matter, its been true for over a century in California. On the eastern side of the Sierras, there was armed conflict in the early years of the 1900s over Los Angeles stealing the water from Owens Valley in what became known as the California Water Wars.

Of the approximately 1% of water present on Earth as liquid fresh water, 96% of that resides below the land surface as ground water, 3.48% is present in lakes (about half salty and half fresh) and swamps, and 0.44% resides in the atmosphere or exists as soil moisture. The rivers of the world hold the rest (0.08%), a grand total of 0.0002% of all of the water on Earth.

Yet it has been this smallest reservoir of H_2O on the planet, this drop in the bucket (or to be more precise, 1/3 of a teaspoon of fresh water in a 2 gallon bucket of the world's total water), which most of the world's people have depended upon for as long as civilization has existed. This is true for the city of Santa Cruz, the state of California, and much of the nation and the world. Whether pumping water directly from rivers as the city of Santa Cruz does from the San Lorenzo River or north coast streams, or taking water from lakes or reservoirs (Loch Lomond on Newell Creek, for example), its almost all river water, whether flowing or dammed.

Throughout California, we built over 500 dams on coastal streams alone throughout the last century, forgetting the other nine hundred in the Sierra Nevada and Central Valley. While these dams were often built for multiple purposes, in virtually every case, water supply was an important benefit. The city of Santa Cruz depends upon streams for about 96% of our water supply, while the state of California typically gets about 60% of its water from rivers, the rest from groundwater.

The streams of the world get their water from rainfall, snowmelt, or groundwater seepage, all of which come ultimately from precipitation. This water is all part of the hydrologic cycle, which starts with evaporation, primarily from the oceans, and which leaves the salt behind. As this moisture or water vapor rises into the atmosphere, it cools and the moisture condenses to form water droplets, or if cold enough, ice crystals. It then falls as precipitation, much of this over the ocean, but also over the continents. This natural distillation or desalting process gives us the supply of fresh water that the lives of the Earth's 7.1 billion people are totally dependent upon. We can live for weeks without food, but only a few days without water.

Article 8

Rainfall, Runoff, and Reservoirs

Loch Lomond reservoir, the major water source for Santa Cruz.
(© Don DeBold licensed through CC BY 2.0 via flickr)

Santa Cruz is one of the few counties in Central or Southern California that are hydrologically self-sufficient; we don't import water through pipelines or tunnels as do many of the other counties, from the San Francisco Bay area to San Diego. Well, unless you count all of the bottled water: Perrier, Fiji, Dasani, Nestle, Mountain Valley, Evian, Aquafina, or any of the other dozens of different brands of bottled water that get trucked over Highway 17.

Other than what comes out of those plastic bottles, we get our water in Santa Cruz County from either what flows across the land surface, or what resides beneath the ground. These two sources globally only make up 1% of all of the water on the planet, and all of the water in either place, streams or groundwater, started out in the ocean as salt water, days or decades before.

Evaporation from the ocean produces fresh water vapor, and on occasion, at least historically, that vapor rises and condenses to fall as precipitation across the county. That rainfall is not distributed uniformly, however, due in large part due to the effects of the Santa Cruz Mountains. As the rain clouds move over the mountains from the coast, they rise, cool, and the moisture condenses. We can therefore expect greater rainfall at higher elevations. Along the coast, for example, Davenport, Santa Cruz, Soquel, and Aptos average between about 26 and 30 inches of rain per year. Bonny Doon and Boulder Creek, on the other hand, typically get closer to 60 inches.

The effect of topography on rainfall is particularly pronounced in two places on Earth that between them hold many of the world's rainfall records. La Reunion, a small Indian Ocean island near Madagascar, is circular in shape with lowlands near the coast but with central peaks reaching over 10,000 feet in elevation. The island lies directly in the path of southwest Indian Ocean tropical cyclones. Among many rainfall intensity records, it has received up to 45 inches of rain in 12 hours, and 71.8 inches of rain (six feet) in 24 hours. That's over twice our entire annual rainfall in just 24 hours!

The other area with stacks of rainfall records is the town of Cherrapunji, India, known as one of the wettest places on Earth. Monsoon clouds from the Bay of Bengal move inland and run smack into the foothills of the Himalayas, dropping all of that moisture on Cherrapunji, where rainfall is measured in feet. This damp little town holds the precipitation record for one month with 31 feet of rain, two months with 42 feet of rain, and a year with 87 feet of rain. They don't generally have water shortages in Cherrapunji, nor do they often leave home without their umbrellas.

When the rain hits the ground anywhere, it basically has two choices. Where the land surface is relatively flat and the soils are permeable, the water can infiltrate or percolate into the subsurface, eventually becoming part of the groundwater system. Where the land surface has more of a slope and/or where the ground is less permeable, perhaps because of bedrock close to the surface, compacted or saturated soils, or the presence of impermeable surfaces like roofs, streets, or parking lots, the water will run off and soon enter a waterway, creek or a stream.

For steep, relatively short watersheds like those of Santa Cruz County (the San Lorenzo River or Soquel Creek, for example), the total time elapsed from when a raindrop hits the ground surface to when that same water drop has flowed the length of the watershed and reaches the ocean (about 30 miles) may only be a day or two. The Amazon River, in contrast, is 4000 miles long, and the time from rainfall to ocean discharge can be weeks.

The City of Santa Cruz water supply system is about 96% dependent on surface water, with stream flow under natural conditions varying seasonally. The great bulk of our rainfall and runoff occurs between November and April, as is the case for much of the west where rainfall dominates over snowfall. This is why the development of the arid southwestern states, and we can count much of California in that category, was accomplished by either damming rivers and creating reservoirs to store that winter runoff, or by diverting or stealing water from someplace that had more of it.

Santa Cruz has only one large reservoir, Loch Lomond, on Newell Creek, near Lompico. The earthfill dam was completed in 1960 and the reservoir has historically provided about 24% of the supply to the city water system.

Article 9

Groundwater - An Invisible Resource

Almost all of the water on Earth (97%) is salty, and the 3% left over is split about one-third / two-thirds between groundwater and ice. There is also a drop in the bucket left over in lakes and rivers (actually just over 0.02%).

Nationwide, groundwater is the source for about 23% of all the fresh water we use, the other 77% comes from surface water such as streams, lakes and reservoirs. In the more arid western states, groundwater provides a much greater proportion of the water supply, in large part because many of the streams just don't provide a reliable year round source. During a drought, the reduction in stream flow is even greater. In California, groundwater has historically provided about 40% of our total water supply.

Where does that groundwater come from? Well, some water, whether from rainfall or stream flow, seeps into the ground, much like pouring a bottle of water onto a pile of sand. The water moves slowly downward through the cracks or voids in the soil, sediment or rock, until it reaches a layer through which it cannot easily flow. The water then will gradually fill up the empty spaces or cracks above that layer. Water that collects within rock or sediment is called groundwater and the upper surface of that layer is called the water table. This is what we would like to hit when we drill a water well. Sometimes we do and sometimes we don't.

Groundwater resides in aquifers, which are geologic materials that groundwater can easily move through. The amount of water that can flow through soil or rock depends upon the size of the empty spaces in the material and how well they are interconnected. Porosity is a measure of the amount of pore space within a rock, and permeability describes how well the spaces are connected. A sponge has both good porosity and permeability; Styrofoam or bubble wrap has lots of pore spaces but they aren't connected.

This reminds me of a story that not only explains porosity and permeability but also provides some life lessons.

A college professor stood before his class at the start of a new semester. Silently, he picked up a very large jar and filled it with golf balls. Then he asked the students if the jar was full. They agreed that it was.

The professor then picked up a box of pebbles and poured them into the jar. He shook the jar lightly, and the pebbles settled into the open areas between the golf balls. He then asked the students again if the jar was full. They again agreed that it was.

The professor next picked up a container of sand and poured it into the jar. He asked once more if the jar was full. The students again responded with a resounding "yes."

The professor then produced a beer from under the table and poured it into the jar, filling the empty spaces between the sand. The students laughed.

"Now," said the professor. "I want you to understand that this jar represents your life. The golf balls are the important things -- your family, friends and health. If everything else was lost and only they remained, your life would still be full."

"The pebbles are the other things that matter - your job, your house, your accomplishments, etc. The sand is everything else - the small stuff. You know, the newest I-Phone, I-Pad or some other gadget."

"If you put the sand into the jar first," he continued, "there's no room left for the golf balls or pebbles. The same holds true for life. If you spend all your time and energy on the small stuff, you'll never have room for the things that are really important."

"Pay attention to the things that are essential to your happiness. Spend time with your friends and family. Sit by the oceans or in the woods. Listen to the birds and smell the flowers. Enjoy the beauty of your existence. There will always be time for the small things. Take care of the golf balls first - the things that really matter. The rest is just sand."

One of the students then raised her hand and asked what the beer represented.

The professor smiled, "I'm glad you asked."

"The beer shows you that, no matter how full your life may seem, there's always room for a beer with a friend."

Article 10

Groundwater
Out of Sight but Not Out of Mind

Our underground aquifers serve as reservoirs, pipelines and filters for the water beneath the surface. As subsurface reservoirs, they have lots of advantages over those at the surface: there are no construction costs; they don't silt up or trap sand destined for our beaches; they don't present seismic hazards as they age; and they don't take up valuable land area.

On the down side, however, water generally moves through aquifers very slowly, in fact, usually incredibly slowly. River velocities may be a foot per second (equivalent to 16 miles in a day), so a spill or pollutant can move downstream and mostly be gone in hours or days. Because groundwater moves so slowly (often a foot per day, or even a foot per year), flushing or removing a pollutant or contaminant from an aquifer can take a very long time.

As a result, regulations in California are very strict regarding what water we can put back into an underground reservoir. We can't just pipe storm-water run-off or recycled wastewater into the ground; it has to be cleaned to a very high level before it can be injected into the subsurface.

This slow motion flow also means that aquifers can take a long time to be refilled when the groundwater table has been lowered. Three years of drought and the continued overdraft of many groundwater basins, led to a historic package of three bills signed by Governor Brown on September 16, 2014, which will for the first time, initiate groundwater sustainability planning and management for California's most distressed and overdrafted aquifers.

We haven't been balancing our groundwater checkbook very well. About 800 billion gallons of water are being drained annually from the Central Valley aquifers. The landmark legislation will require local governments to bring groundwater basins up to sustainable levels and limit future withdrawals to the rate of natural replenishment.

An additional problem with heavy groundwater usage, well known in the Santa Clara and Central valleys, is the subsidence of the ground surface accompanying the drawdown of the water table. In the 1950s and 1960s, tract homes in the Santa Clara Valley replaced fruit orchards and water demand increased dramatically.

As ground water pumping increased, the water table dropped as much as 250 feet. This led to compaction of the sediments in the aquifers and the settlement of the ground surface by as much as 13 feet in downtown San Jose by 1967. San Jose had the dubious distinction of being the first area in the United States where land subsidence was recognized as having been caused solely by excessive ground water removal.

While this may not seem like such a big deal, ground settlement produced major and expensive problems: sewage lines, which flow by gravity, no longer sloped in the right direction; a lowered land surface around the nearly flat-lying shoreline of southern San Francisco Bay led to flooding at high tides and the need to construct levees; streams overflowed more frequently; and water wells had to be repaired or redrilled.

The world's largest area of intense land subsidence from ground water withdrawal occurred in the San Joaquin Valley where over 5,000 square miles of agriculture land was affected. Maximum ground surface subsidence, near Mendota, reached over 28 feet. This produced severe problems in the construction and maintenance of water transport structures, such as canals, irrigation and drainage systems. In both the Santa Clara and San Joaquin valleys, subsidence has been terminated for the most part by importing surface water, and allowing the water table to begin to recover.

Excess pumping or overdraft of aquifers along coastlines creates another set of problems. In coastal areas, fresh water aquifers are commonly in contact with the ocean and are exposed to seawater. As long as the water table within the aquifer is above sea level, the inflow of salt water is repelled and little or no contamination occurs.

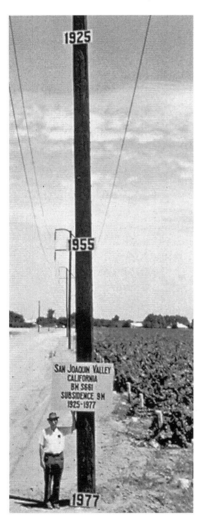

48 years of ground subsidence. (courtesy: US Geological Survey)

However, the increased demand for groundwater in many fertile coastal plain areas (the lower Pajaro and Salinas Valley areas, for example, but there are many, many others) has led to the lowering of the water table below sea level and the intrusion of a wedge of seawater into aquifers. Salt or brackish water soon appears in wells, making the water undrinkable and unusable for crops.

Since the 1940s, continuous heavy pumping of groundwater from the lower Salinas Valley between Moss Landing and Marina lowered the ground water table and allowed seawater to intrude the upper 180-foot aquifer. Wells were abandoned as they began to pump brackish water and new wells were drilled farther inland. The saltwater plume continued to move inland, and by 2011 it had reached nearly to Salinas, over 8 miles from the shoreline.

Article 11

Droughts and Mega-Droughts

In July I was comparing water-rationing rules with a friend from Manhattan Beach, wondering which city had the most stringent limitations. Out of curiosity I asked "How much rain did you get this past winter?" It took a few seconds for him to answer, and then with a grim look on his face he replied with, "It didn't rain in Manhattan Beach this winter".

The local rainfall in the greater Los Angeles area actually has little impact on their overall water supply, however, as it all comes from somewhere else, and has for decades. Los Angeles has always been a desert, masquerading as a tropical island with their lush gardens, lawns and golf courses, all thanks to imported water.

But it did make me feel a little better for an instant, living in Santa Cruz, where we recorded a grand total of 13.27 inches of rain this year. That was the glass half-full reaction; the half-empty side of the picture is that it's the 3rd year in a row of significantly below average precipitation, and has been one of the driest years, if not the driest, since record keeping began here 150 years ago. 2014 is also on track to also be the warmest year on record in California.

The definition of drought is "a prolonged period of abnormally low rainfall". A new term has been added to our weather conversations in recent years, megadrought, defined as "a prolonged drought lasting two decades or longer". This is a word most water suppliers would rather not think about.

We are now in the midst of what is being labeled as the worst drought in the western U.S. since we started keeping track of these events about 150 years ago. The parched area now covers a dozen states and nearly 600 counties, from southern Texas to the northern

Rockies, and includes grazing and farm land that produce a third of the country's beef cattle and half of its fruit, vegetables and winter wheat.

California has been hit the hardest, however. As of October, 2014, nearly 60 percent of the state is officially in an "exceptional" drought - the highest level of aridity, with 82% of the state in the next driest state, "extreme" - and those folks who watch the weather for a living aren't seeing any clouds with silver linings in the near future.

Not surprising to long-time California residents, droughts are not new or uncommon. We get these dry periods every decade or two. The last local drought extended from 1996 to 2001, when Santa Cruz averaged just 19.5 inches annually for those 5 years, compared to our long-term average of just over 30 inches. The 1975-1976 drought was even drier but shorter, averaging less than 15 inches per year. For the past 3 years, we averaged 17.6 inches.

If you really want to get concerned and lose sleep, the pre-historic record has preserved far more serious droughts in the distant past, megadroughts in today's terminology. But this was before California was home to 38 million people and our farms and fields were providing fruits, vegetables and livestock to the nation. In the early years of the last century, if there wasn't enough water where we wanted a city or farm, we built dams, canals and pipelines to move the state's water to where we needed it. And more often than not, we fought over who had the rights to the water. Between 1860 and 2000, 1400 larger dams (over 25 feet high) were built on California's rivers; that's an average of 10 per year or almost one every month for 140 years.

Article 12

Reading Tree Rings

Cross section of a Blue Spruce showing variable ring widths.
© 2014 Gary Griggs

Droughts over the past two centuries in California have usually lasted several years at most, and there are many accounts of how the dry periods in the 1800s affected livestock, agriculture, and people. Until the last decade or so of the 1800s, however, there were no reservoirs to store winter runoff, so low rainfall years hit particularly hard on the early settlers. Were the last 200 years typical? That's an important question to try and answer today.

Rainfall records in the state only go back about 150 years, but dendrochronology, or the study of tree rings, has allowed us to look much further back in time. Just like putting on a few extra pounds around your waist when you eat well on a vacation, trees suck up moisture when there is plenty to go around, and use that extra water to grow thicker rings.

History has been written in many places other than in books, and the job of a paleoclimatologist is to find where those climate records have been preserved. Tree rings, lake and seafloor sediments, corals, and ice are a few places where we have been successful in extracting climate records.

Bristlecone pines, which can live to be 4000-5000 years old, and survive in the White Mountains of southeastern California, are living history books in which the records of our pre-historic rainfall have been preserved. These ancient trees contain the evidence in the width of their annual rings that the last few thousand years have been characterized by alternating 50 to 90-year wet and dry periods, but also by droughts that have lasted 10-20 years.

These are mild events, however, compared to the period from about 900 to 1400 A.D., known globally as the Medieval Warm Period. Evidence from detailed tree ring studies indicates that during this period, droughts as long as a century were common.

Is the present three-year period of well below average precipitation just another short-term drought, or is this the beginning of another megadrought, perhaps intensified by global climate change? No one knows, yet. But most climate scientists would agree that the past century was unusually wet, and it has been during the past 100 years that the populations of the arid southwestern states literally exploded. Nevada went from 42,000 people in 1900 to 2.7 million today. Arizona grew from 123,000 to 6.4 million during the same period. And these states are both deserts.

California has 38 million people today, 25 times more than in 1900. And all this growth, whether farms, factories or cities, took place during what was very likely an unusually wet century.

For decades, Central Valley agriculture had everything- sunshine, fertile soils, and water. But that water was imported from elsewhere or pumped from aquifers that were being depleted year after year. This year, 430,000 acres of productive farmland in California were left fallow due to the drought. As of late August, 95% of the entire state was designated as being under severe drought conditions, and most of our largest reservoirs were at just one-quarter to one-third of their total capacity.

If we are entering a longer-term drought, the impacts and implications for California are enormous. It's hard to believe this is possible, but history tells us it has happened in the past. We have lived our entire lives in times of abundant water, but we need to begin facing the potential of a different future.

We need to become even more efficient in our use of water, especially in our agricultural practices, but also in our cities. While there are no technical difficulties with recycling our wastewater, there are still regulations that prohibit its usage except for certain types of irrigation. Without significant rainfall, it's difficult to capture additional runoff. So where does that leave us?

Article 12

Water-Searching for Answers

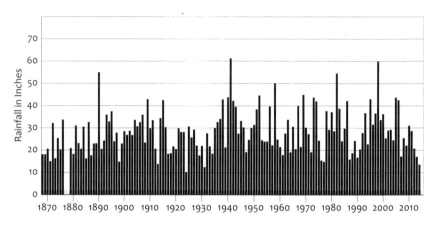

Rainfall for Santa Cruz from 1867 to 2014
(© D Shrestha Ross, based on data from Western Regional Climate Center and the National Weather Service)

We can live without food for 3 weeks or so, without air for about 3 minutes, and without water for 3 days. The amount of water on Earth doesn't change, but the population needing that water is increasing daily by about 200,000 people.

Nearly 97% of all the water on Earth is in the oceans and is salty. That percentage is gradually increasing, however, as ice sheets continue to melt and glaciers retreat. Of the roughly 3% of the water on Earth that is fresh, about ½ of it is wrapped up in those ice sheets and glaciers and isn't readily accessible.

We just experienced one of the California's driest years, in fact the 3rd driest in 119 years of record keeping, after two prior very dry years. What's coming next? To be completely honest, we don't know. We do know that the southwestern U.S. has experienced droughts that have lasted a decade or two, and even longer. We also know that the Earth is warming up and this isn't likely to stop any time soon. This doesn't help our water supply situation, and will most likely make things worse.

On the positive side, Santa Cruz has done exceptionally well with water conservation during this year's drought. In September we ranked among the state's communities with the lowest per person water usage at 49 gallons/day. We all deserve an A+ for water conservation.

Another group of California communities and water districts gets an "F" in water usage in my grade book. They used more than 10 times as much water per day as we did, and they are all in Orange and San Diego Counties, some of the hottest and driest parts of the state.

Most of us understand our present situation and are willing to do our part. How long we can keep this up, though? What if we have a 4th dry year, or a 5th? Or a decade of low rainfall? How long are we all willing to live with dying plants and brown lawns, wilting vegetable gardens, short showers and dirty cars? Or do we have a choice?

About 96% of our city water supply comes from runoff, which depends directly on rainfall, and that's been a little scarce lately. We can store some of that runoff, but our only large storage tank, Loch Lomond, is now at 59% of capacity and its lowest level in 23 years.

There is an important process underway now with the establishment of the City's Water Supply Advisory Committee and their public meetings and open discussions. There are opportunities for lots of input, and interested citizens haven't been hesitant or shy in suggesting a lot of creative ideas.

We are a small county with a limited amount of rainfall, and therefore, a limited water supply. There isn't going to be a pipeline delivering water to us from somewhere else anytime soon. So we need to be creative, real creative, and also not be too quick to discard any ideas.

There are basically two options, reduce our usage or increase our supply, or a combination of the two. So far, it's been all about reducing usage and we've done really well. But how much more we can conserve and is this sustainable?

A very distinguished and prominent civil engineer at UC Berkeley once said to me: "For every complex problem there is always a simple answer, and it's always wrong."

Any solution (or set of solutions) is going to cost money, and our future water is definitely going to cost significantly more than what we've paid in the past-simply because there hasn't been a big investment in new sources, new storage, or conveyance, for many years. Every proposal will also have environmental impacts that will have to be thoughtfully evaluated, and any project will take time to plan, fund and construct.

And we can only live 3 to 4 days without water.

Made in the USA
San Bernardino, CA
05 December 2014